Regional Development in China

Regional inequality is an issue of concern for many countries, creating problems of national disunity and social instability. In China, the issue has been of particular importance, as a result of its role in debates concerning the nature of socialism, central control, and local autonomy. This book systematically examines uneven regional development in China, and places it in a global and theoretical context.

The book illuminates our understanding of the issue in three key areas:

- An exhaustive case study of Jiangsu province in East China provides a wealth of new empirical source material.
- By explaining uneven development in terms of the interaction of three central agents (the foreign investor, the state and the region), the book opens up a new theoretical approach to the subject.
- The author's conclusions have important implications for policy, and will be of vital interest for governments and businesses involved in China.

Regional Development in China is a vital source of reference for those with an interest in transition economies, and makes a substantial contribution to the study of globalization and liberalization in the developing world.

Yehua Dennis Wei is Assistant Professor of Geography at the University of Wisconsin-Milwaukee (UWM), and is affiliated to the Urban Studies Program, the Asian Studies Program and the Center for International Studies at UWM. He has published widely on China's regional and urban development, urbanization, population, and political economy.

Routledge Studies on China in Transition
Series Editor: David S. G. Goodman

1 The Democratisation of China
Baogang He

2 Beyond Beijing
Dali Yang

3 China's Enterprise Reform
Changing state/society relations after Mao
You Ji

4 Industrial Change in China
Economic restructuring and conflicting interests
Kate Hannan

5 The Entrepreneurial State in China
Real estate and commerce departments in reform era Tianjin
Jane Duckett

6 Tourism and Modernity in China
Tim Oakes

7 Cities in Post Mao China
Recipes for Economic Development in the Reform Era
Jae Ho Chung

8 China's Spatial Economic Development
Regional Transformation in the Lower Yangzi Delta
Andrew M Marton

9 Regional Development in China
States, Globalization, and Inequality
Yehua Dennis Wei

Regional Development in China

States, Globalization, and Inequality

Yehua Dennis Wei

ROUTLEDGE
Taylor & Francis Group

London and New York

First published 2000
by Routledge
11 New Fetter Lane, London EC4P 4EE

Simultaneously published in the USA and Canada
by Routledge
29 West 35th Street, New York, NY 10001

Routledge is an imprint of the Taylor & Francis Group

Typeset in 10/12 pt Garamond by
Newgen Imaging Systems (P) Ltd., Chennai, India
Printed and bound in Great Britain by TJ International Ltd, Padstow, Cornwall

British Library Cataloguing in Publication Data
A catalogue record for this book is available
from the British Library

Library of Congress Cataloging in Publication Data

Wei, Yehua Dennis, 1963–
Regional development in China: states, globalization, and inequality/Yehua Dennis Wei.
p. cm.
Includes bibliographical references and index.
1. China—Economic conditions—1976—Regional disparities. 2. Regional disparities—China. 3. China—Economic policy—1976– I. Title.

HC427.92.W436 2000
338.951—dc21 00-032309

ISBN 0-415-22448-9

Contents

List of figures

List of tables

Acknowledgements

I am indebted to many friends, colleagues, and teachers. I would like especially to thank Larry Ma and Cindy Fan for their careful guidance and constant support over the years. Many commented on or provided materials for my research, including Ashok Dutt, Jams Tong, Richard Baum, Andy Wedeman, and Dali Yang in the United States, and Hu Xuwei, Lu Dadao, Zheng Weiyuan, Yang Kaizhong, Wei Houkai, and He Denggao in China. At the University of Wisconsin at Milwaukee (UWM), I would like to express my gratitude to David Buck, Sunny Kim, Donna Schenström at the Cartographic Lab, and Ruihong Huang and Michelle Davis for research assistance. My appreciation also goes to David S.G. Goodman, whose expertise in regional development makes the series ideal for such a project.

I would like to acknowledge the financial support from UWM and UCLA. At UWM, my research and fieldwork were funded by the Graduate School Research Committee award, Department of Geography, Joint Center for International Studies, College of Letters and Sciences, and Office of International Studies. I am also grateful for UCLA's International Studies and Overseas program for funding my fieldwork in China.

I appreciate the following journals for granting me permission to draw on previously published materials: 'Fiscal Systems and Uneven Regional Development in China, 1978–1991,' *Geoforum* 27(3) (1996), 329–344; 'Determinants of State Investment in China, 1953–1990,' *Tijdschrift voor Economische en Sociale Geografie* (*Journal of Economic and Social Geography*), 88(3) (1997), 211–225 (with Larry Ma); 'Changes in Industrial Ownership Structure in China,' *Geography* 84(3) (1999): 193–203. Finally, I would like to thank my wife, Weimei Ye, for her support during my research.

Abbreviations

CCISOE	capital construction investment in state-owned enterprises
CCP	Chinese Communist Party
CDS	coastal development strategy
CJV	contractual joint venture
COE	collective-owned enterprise
CV	coefficient of variation
EJV	equity joint venture
FAI	fixed asset investment
FDI	foreign direct investment
FIE	foreign-invested enterprise
FYP	Five-Year Plan
GDP	gross domestic product
GDPP	gross domestic product per capita
GLF	Great Leap Forward (1958–60)
GVAO	gross value of agricultural output
GVIAO	gross value of industrial and agricultural output
GVIAOP	gross value of industrial and agricultural output per capita
GVIO	gross value of industrial output
IOE	individual-owned enterprise
JE	joint exploration
JSB	Jiangsu Statistical Bureau
LQ	location quotient
MNE	multinational enterprise
MOF	Ministry of Finance
NIDL	New International Division of Labour
NI	national income (net material product)
NIP	national income per capita
NIC	newly industrializing country
OCC	open coastal city
PBC	People's Bank of China
PRC	People's Republic of China
SOE	state-owned enterprise
SSB	State Statistical Bureau

SEZ	special economic zone
SHE	shareholding enterprise
SPC	state planning commission
TE	township enterprise
TNC	transnational corporation
TVE	township and village enterprise
VE	village enterprise
WFOE	wholly foreign-owned enterprise

1 Introduction

Regional development is an important area of academic inquiry. Regional inequality in particular is one of the major concerns facing governments, as it reflects unequal opportunities among regions and may threaten national unity and social stability (Smith 1995; World Bank 1997).[1] Regional development and inequality are of particular importance to the developing countries that inherited regional imbalances as a result of colonialism, and to former socialist countries that demanded, as a point of ideology, a reduction in regional inequality. Even in developed countries, such as the United Kingdom, Italy, and Canada, regional inequality has received considerable attention (Martin 1988; Higgins and Savoie 1995). The trends of and forces underlying regional inequality have been the subject of heated debate. Questions whether regional inequality was reduced in former socialist countries and whether economic reforms have intensified regional inequality have generated considerable research. Indeed, spatial issues are the traditional areas of interest, especially for geographers and planners, but have attracted broad interests in economics, geography, political science, sociology, and history (e.g., Parsons 1988; Storper and Walker 1989; Porter 1990; Krugman 1991, 1995; Plestina 1992).

Regional inequality has been an important issue in China since the establishment of the People's Republic of China (PRC) in 1949, mainly as a result of its role in debates concerning the nature of socialism, central control and local autonomy, and resource allocation among regions (Bo 1993). Researchers generally agree that some effort was expended to develop economically lagging regions but there is disagreement over the extent and consequences of this (Wei 1999a). Since the launch of the economic reforms in 1978, China's dominant development policies have shifted from ones based on self-reliance to ones based on comparative advantage and open door policy. Problems arising from migration, resource allocation, and intense intergovernmental relations have prompted many to consider regional inequality as an important issue of government policy, although researchers disagree over whether regional inequality has in fact intensified and what forces have contributed to changing patterns of regional development.

This book investigates changing patterns and underlying factors of regional development and inequality in China since the founding of the PRC, with an emphasis on the reform period. I adopt a multiscalar approach to examine changing patterns of regional inequality, and argue for an analysis of the central state, local states and factors, and global forces to achieve a better understanding of China's regional development. I hope this study improves the research on the trends and causal mechanisms of regional inequality in China and contributes to the broad debates on theories and policies of regional development and the transition taking place in the former socialist countries.

1.1 China's development path and regions

When the PRC was established, Chairman Mao Zedong and the Chinese Communist Party (CCP) faced the problems of poverty and substantial coastal–interior imbalances in China. Influenced by socialist ideology and egalitarian ideas, Mao tried to develop the backward inland areas and to reduce regional inequality. Indeed, the appropriate allocation of resources among regions was one of the major components of China's First Five-Year Plan (FYP) (1953–57), and has since become an important part of China's national and regional planning. The relationship between the coastal and interior regions was one of the 'ten great relationships' analyzed by Mao (1956) in which he stressed the development of the interior region while encouraging growth in the coastal region. Researchers generally agree that some effort was made to develop the interior, but there is disagreement over the extent and consequences of such a policy. Most maintain that regional inequality declined as a result of the implementation of distributive policies, while others contend that regional inequality actually persisted or even increased under Mao (e.g., Lardy 1975, 1978, 1980; Donnithorne 1976; Paine 1981; Cole 1987; Wu 1987; Pannell 1988; Cannon 1990; Yang 1990; Lyons 1991; Tsui 1991; Wei and Ma 1996; Wei 1998).

Since the late 1970s, China has experienced dramatic reforms and profound changes. Influenced by the 'ladder-step theory' (*tidu lilun*)—a Chinese version of the growth pole and inverted-U theories—the government has encouraged some regions to 'get rich quick', thereby emphasizing the development of the coastal region. It argues that concentration and specialization are required to accelerate economic growth in China and that spillover from the more developed coastal region would stimulate the prosperity of the country as a whole (State Council 1986). China's reform policies in the 1980s, such as the open door policy and the coastal development strategy, provided more favorable policies to some coastal areas. Supporters of uneven development policies argue that these policies have stimulated rapid economic growth without generating huge regional gaps, while the opposition contends that social problems arising from reforms, including rising regional inequality, have brought social unrest to

China. To what extent regional inequality has increased in post-Mao China and how many resources should be committed to developing the backward regions have become central issues of government policy and research agenda.

After a short period of stagnation following the Tiananmen incident in 1989, the 1990s witnessed the implementation of bolder reforms aimed toward transforming China into a 'socialist market economy.' During the 1990s, some regions, particularly some coastal areas such as southern Jiangsu, recorded dramatic growth while some interior areas were lagging far behind. Millions of rural workers migrated to coastal cities in pursuit of employment and better living conditions, which generated considerable concerns about the impact of economic reforms and spatial policies in China. Regional 'resource wars' erupted in many regions in which regions used illegal administrative and even military measures to protect their local markets and restrict interregional resource flows.[2] Bargaining between central and local governments over resource allocation and development policies became extremely intense, partly as a result of China's poorer areas considering themselves to be less favored by the central government. These problems—intensified by inflation and corruption—have prompted many to consider regional inequality as the root of China's regional problems and an important issue in government policy. The Ninth FYP (1996–2000) views polarization as a serious threat to China's future prosperity, stability, and unity; consequently, the government has made reducing regional inequality a top policy priority.

Researchers disagree once more over whether regional inequality has intensified and over the forces that have contributed to the change of regional inequality in post-Mao China (e.g., Lakshmanan and Hua 1987; Wu 1987; Aguignier 1988; Goodman 1989; Kueh 1989; Cannon 1990; Linge and Forbes 1990; Denny 1991; Lyons 1991; Tsui 1991; Veeck 1991a; Kato 1992; Huo 1994; Beresford and McFarlane 1995; Fan 1995a; Chai 1996; Chen and Fleisher 1996; Jian *et al.* 1996; Wei and Ma 1996; Zhao 1996; Yang 1997; Raiser 1998). The debate is also intense among scholars in China (e.g., Yang 1992, 1993; Hu *et al.* 1995; H. Wei 1995; Lu and Sit 1997; Wang *et al.* 1998). The discourse surrounding China's regional inequality, however, suffers from several shortcomings. First, intellectual and political debates have been principally conducted without concrete evidence, due in part to the lack of reliable time series regional data on social and economic development. Secondly, even fewer studies have been undertaken to explain the forces responsible for regional inequality. Thirdly, the scale of observation has created much confusion, as the patterns of change across the nation's macro geographic regions (coastal versus interior regions or eastern, central, and western regions) are different from those across the provinces (Wei and Ma 1996; Zhao 1996). Lastly, substantial regional inequalities exist within the provinces, and merit further investigation (Wei and Fan 1997; Yang 1997; Lyons 1998).

1.2 Objectives of the research

In this research, I investigate regional development from the founding of the PRC to the late 1990s, with an emphasis on regional inequality during the reform period. I attempt to systematically examine disparities in reform policies and development processes across China's regions. The first objective is to examine changing patterns of regional development, and this in turn will determine whether, and to what extent, regional inequality has intensified. This research uses a multiscalar approach, as I investigate regional inequality across China's macro regions, provinces, and counties (interregional, interprovincial, and intraprovincial inequalities). This approach is sensitive to geographic scales and shows the complex mosaic of regional development emerging in China. The second objective is to analyze the impact of three major forces—the central state, local states and agents, and global forces—on regional inequality.[3] Regional development theories are developed largely based on the assumptions of closed free market economies and the experiences of developed countries, and are preoccupied with a singular focus explanation of regional inequality. Each theory only captures, at best, part of the picture of regional inequality. They are less relevant for explaining regional inequality in transition economies, such as China, where forces of regional development differ from that in capitalist economies. This research is within a multivariate framework and explores how the complex interplay of national, local, and global forces shapes regional development. Thirdly, the new findings will also provide policy and theoretical implications for regional development and poverty relief. The analysis of China may provide useful references for other transition economies and developing countries that are undergoing similar processes of economic restructuring and globalization.

I argue for multiscale and multi-mechanism approaches for a better understanding of regional inequality in China. First, I maintain that three particularly powerful factors and agents underlie regional development in China: namely, the central state, the local state and agent, and the foreign investor. Scholars have attempted to understand economic reforms and transition taking place in China (e.g., Nee 1989; Oi 1992; Naughton 1995a; Walder 1995a), and the spatial ramification of reforms (e.g., Fan 1995a; Lin 1997; Yang 1997; Wei 1999a). Factors and agents of regional development must be analyzed in light of the complexities of the Chinese economy. In China, the central state is still playing an important role, and global and local forces have emerged as important forces shaping China's regional development, as a consequence of globalization and China's economic reforms (Wei 1997a; Wei and Fan 1997). I wish to show that the multitude of agents has jointly produced uneven regional development in China.

Secondly, I investigate the regions and adopt a multiscalar approach. An analysis of regions, including development paths and assets, helps to unfold how the above three sets of underlying forces shape changing patterns of regional development. In addition, regional analysis relates to geographic scale (Gore 1984). Patterns of regional inequality across China's macro regions differ from that across the provinces. Within provinces, inequalities across regions also differ from that across the counties. In other words, a multiscale approach is helpful when studying regional inequality, and we should not exaggerate or ignore the significance of any level of regional inequality. This book examines regional inequality across China's macro regions (coastal, central, and western regions), across provinces, and within provinces (through a case study of Jiangsu), and analyzes development paths and conditions of selected regions.

Thirdly, the temporal change of regional inequality cannot be simply described as convergent, divergent, or inverted-U. Rather, it is shaped by multi-mechanisms, related to the fortune of regions, and is specific to geographic scale. The forces underlying regional development change over time, and the impact of the global, state, and local forces on regional inequality differs historically and geographically. To better understand these three arguments, I briefly review the literature on regional development and the research on former socialist countries in the following two sections.

1.3 Regional development theories and the changing world

The issue of regional inequality has been a subject of intense theoretical debates, especially since the 1950s. Scholarly views on the temporal patterns of regional inequality, the forces driving regional development, and the roles of institutions differ greatly. Neoclassical economics, growth pole dynamics, and inverted-U theories maintain that factor mobility and diffusion tend to equalize regional income differentials, while divergence and structural theories argue that backwash effects and capital accumulation tend to reinforce regional inequality. Theoretical pluralism and debates characterize theories of regional inequality.

The neoclassical growth theory emphasizes equilibrium conditions and the importance of the market in allocating resources, and considers regional inequality as a transitory phenomenon. It stresses the influence of the supply-side factors such as labor, capital stock, and technical change, and views regional growth as a process of resource reallocation, i.e. the mobility of capital and labor. It holds that regional inequality rises from temporary disequilibrium between supply and demand; efficient markets and factor mobility would eventually equalize regional differentials (Borts and Stein 1964; Smith 1975). Regional export base theory also predicts long-term regional convergence (North 1955). The neoclassical growth theory has been influential in studying regional inequality, but has been challenged on several grounds, especially by institutionalists and structuralists.

Growth pole and inverted-U theories are probably the most influential in regional development and planning in developing countries. Perroux (1950, 1955) considers growth poles as engines for innovation and regional growth.

The inverted-U theory holds that regional inequality tends to rise during the early stages of development and to fall as the economy matures (e.g., Hirschman 1958; Williamson 1965; Friedmann 1966; Alonso 1980). This paradigm of regional development is also known as top-down development, or development from above (Hansen 1981). Since the late 1960s, a great deal of empirical work has been conducted. Some scholars have supported the convergence and inverted-U theories (e.g., Green 1969; Mera 1977; Tabuchi 1988; Mallick 1993; Crown and Wheat 1995; Barro and Sala-i-Martin 1995). Others have criticized these theories and have found regional inequality persisted or even increased in some countries (e.g., Gilbert and Goodman 1976; Stohr and Todtling 1977; Fields and Schultz 1980; Krebs 1982; Maxwell and Hite 1992; Siriopoulos and Asteriou 1998). Development from below and a variety of structural theories appeared as alternatives (Stohr and Taylor 1981).

The divergence theories, particularly dependency and structural schools, contend that regional inequality and its increase are inevitable under capitalism. Myrdal's (1957) cumulative causation thesis argues that factors of production tend to move to the more developed core regions, and this negative backwash effect reinforces regional inequality, although he recognizes the work of spread effects. The dependency school holds that the spread of capital to the periphery creates a dependent core–peripheral structure, sustains labor outflow and unequal transfer of value, exacerbates the stagnation of the periphery, and intensifies regional inequality (Frank 1967; Emmanuel 1972; Slater 1975). Regional inequality under capitalism has been seen as necessary for capital accumulation, and the capitalist mode of production actively creates, intensifies, and maintains regional inequality, as capital tends to move to core areas with higher profit rates (Soja 1980; Harvey 1982). Scholars have also stressed the periodic crises of the capitalist economy and the dynamics of uneven regional development (e.g., Smith 1984; Clark *et al.* 1986; Markusen 1987). It is argued that capital tends to move back and forth between developed and underdeveloped areas, and regions are continuously developed and abandoned. Based on the long wave theory, the regional cycle theory argues that regional economies progress in a wave-like manner, causing regional inequality to rise and fall cyclically (Amos 1990).

Since the mid-1980s, considerable attention has been drawn to neo-Marshallian and transactions cost theories of regional development. Scholars argue that Fordist mass production is formed geographically in manufacturing regions and stimulates urbanization and intensifies regional inequality, while post-Fordist production is characterized by flexible production, reinforcing agglomeration, and the development of industrial districts (e.g., Piore and Sabel 1984; Scott 1988). However, these theories have also received numerous critiques for dualism between Fordism and post-Fordism, for inadequate attention to the role of the nation-state, and for poor validity.

Two concepts of convergence have been identified and studied:

1. σ (sigma)-convergence, the most often studied, occurs when the dispersion of per capita income or outputs across regions declines over time;
2. β (beta)-convergence refers to the tendency for initially poorer regions to grow more rapidly than the richer regions (Barro and Sala-i-Martin 1991, 1992, 1995).

Recent work by economists has advocated the notions of increasing returns, agglomeration, and the cumulative causation process (e.g., Krugman 1991, 1995; Arthur 1994). Moreover, endogenous growth theory argues that countries rich in capital, highly educated, and favorable to knowledge accumulation through learning by doing or technological innovations tend to grow faster (e.g., Romer 1986; Lucas 1988; Mankiw *et al.* 1992). The endogenous growth theory, however, has major limitations when applied to a regional context (Martin and Sunley 1998).

Three additional developments are relevant to my study of regional inequality in China. First, geography and history matter. An increasing interest in the difference that space makes has emerged since the mid-1980s (Gregory and Urry 1985). Massey argues that 'the fact that processes take place over space, the facts of distance, of closeness, of geographical variation between areas, of the individual character and meaning of specific places and regions—all these are essential to the operation of social processes themselves' (Massey 1985, p. 52). Spatial processes and patterns vary across time and space, and make a difference to broad social, economic, and political processes.

Secondly, an emerging body of literature has argued for the importance of institutions, especially states and networks (Granovetter 1985; Rich and Linge 1991; Appelbaum and Henderson 1992). States, particularly in Asia, have played a critical role in economic and regional development, although scholars debate over the nature of the state and the implications of the Asian financial crisis. Moreover, scholars have argued for the importance of territorially grounded networks and development paths in regional development (e.g., Lee and Wills 1997; Storper 1997).

Finally, globalization has become an important dimension of our society, and global forces have become influential in spatial development (Dicken 1998). While some people have speculated on the homogenization of the world, the dominance of global cities, and the emergence of 'global villages', others have found the survival of localism, the localization of the global, and the emergence of new industrial districts (e.g., Friedmann 1986; Amin and Thrift 1994; Knox and Taylor 1995). Evans (1979) emphasized the alliance of multinational, state, and local capital in shaping Third World development. It has been suggested that three mechanisms—world economy, nation-state, and locality—are critical to regional and geographic inquiry (Taylor 1993; Agnew and Corbridge 1995).[4]

1.4 Regional development in former socialist countries

Relatively less research has been done on regional development in (former) socialist countries, despite dramatic restructuring in these countries and increasing global interactions. The literature of regional development would not be complete without the investigation of these countries.

It is well known that socialist ideology mandates for the reduction of social and spatial inequalities, particularly the three major differences existing between town and countryside, industry and agriculture, and manual and mental labor. According to the orthodox Marxist view, in all capitalist nations, contradictions and conflicts exist in these three sets of social elements. Under such a scenario, industries are located in the cities and owned by the urban capitalists. The capitalists perform mental work and they always exploit the rural people performing manual labor in the countryside. The gap between city and country, industry and agriculture, and mental and physical labor can be eliminated under socialism when agriculture is transformed into state-owned enterprises, when the productive forces in the city and the country are fully developed, and when the education of the laboring masses is raised to the same level as the urban dwellers (Marx and Engels 1959, 1976; Engels 1975). In the views of Marx and Engels, all people and regions in a nation should have equal shares of resources and development. The less the inequality, the more the country can be viewed as developed. Regional imbalance is considered as a problem to be corrected by policies toward a more balanced pattern of development. Such a view has had considerable impacts on spatial planning in socialist countries.

The equalization of regional development was often seen as an important hallmark of socialism, and one that can be achieved only under socialism (Koropeckyj 1972; Forbes and Thrift 1987; Sidaway and Simon 1990). Some scholars argue that socialist countries followed socialist ideologies of equality and egalitarianism, and their achievements in poverty relief and basic needs satisfaction were remarkable. They held that industrialization in socialist countries was not accompanied by urbanization and metropolitan growth, due partly to policies to control migration and urbanization; they argue that socialist development policies had prompted rational population distribution and equal regional development (e.g., Murphey 1976; Simmons 1979; Chatterjee 1989).

Other scholars, however, take a pragmatic position and have revealed many serious problems in socialist countries. Fuchs and Demko's (1979) review of socialist countries leads them to conclude that after decades of socialist construction, these countries still exhibited marked regional and urban–rural inequalities. Clark (1983) and Liebowitz (1987) also point out that, while the Soviet Union appeared to have achieved some progress in reducing regional inequality, significant developmental gaps remained. Schiffer (1985) argues that inter-republican inequalities in the former

Soviet Union increased throughout the postwar period, and suggests that the critical force affecting regional development is a political concern for efficiency over equality in investment allocation, and regional equality was never the dominating objective of development. Thrift and Forbes' (1986) study of Vietnam reveals the importance of warfare and national security in spatial development. Moreover, the formation of control hierarchy, the development of territorial production complexes, and the incremental approach of central planning had all contributed to the maintenance of the economic advantages of the cores (Turnock 1984). The Soviet strategy of balanced growth failed to equalize regional development (Ozornoy 1991).

Research on socialist countries is often handicapped by a lack of data and information, which has led to many misconceptions about those countries. In fact, socialist governments did not always follow the socialist ideology manifested by Marx and Engels. Development policies in socialist countries were often inconsistent, largely determined by the leaders' interpretation of Marx's theory and their perception of international and domestic conditions. Szelenyi (1981, p. 12) argues that the examination of socialist countries 'cannot be based on ideological speculation, but has to rely on the empirical analysis of the "real processes of urban and regional development."' The process of regional development in socialist countries remains an area subject to further research.

Former socialist countries have taken drastically different approaches to economic reforms. In contrast to China's gradual approaches, East European countries have emphasized 'shock therapy,' with privatization as the defining strategy for transition from socialism to capitalism. Underlying this strategy are the views that free markets and central planning are mutually exclusive and that radical reforms are necessary for the transition to capitalism. However, Russia's economic reforms were marked by economic depression and high social costs. There exist sharp disagreements regarding the process and institutional frameworks of transition (e.g., Nee and Stark 1989; Nolan 1995; Hausner *et al.* 1995; Woo *et al.* 1997).

Meanwhile, scholars are widely concerned that economic reforms in transitional countries have intensified regional inequality. This concern has had tremendous influence in these countries as it highlights the battle over development paths and power struggles among decision makers. Regional inequality is also an issue that is related to ethnic relations, social stability, and political unity. The breakup of the former Yugoslavia and the former Soviet Union is at least partially related to perceived large inequalities among different ethnic groups and regions (Plestina 1992). However, scholars once again disagree over the impact of economic reforms on regional inequality. While some argue for a decline of inequalities due to problems with traditional heavy industrial bases, others suggest a trend of regional divergence and fragmentation (e.g., Ozornoy 1991; Csefalvay 1994; Nemes Nagy 1994; Smith 1998). It is generally accepted that regions are undergoing uneven processes of restructuring and transition, during which there is the emergence of winners and losers, but it is less clear-cut as to what determines changing patterns of regional development and whether the process of regional transition is path-dependent (Pickles and Smith 1998). Compared to Western societies, fewer scholars have examined the role of the state and foreign investment in regional development in transition economies. China is such a country that has been experiencing dramatic restructuring and its regional inequality has raised considerable concerns among scholars and governments.

1.5 Research methods and the Jiangsu case

The release of regional statistics by the Chinese government has provided an incentive for studies on regional development in China. I use national and regional data reported by the State Statistical Bureau (SSB) of China, especially SSB (1990–98), which include comprehensive historical data on China's economic, social, and demographic changes from 1949 to 1997 for all of the nation's provinces. The release of this data has made it possible to comprehensively examine regional inequality in China for the last five decades. In terms of subprovincial data, Jiangsu Statistics Bureau (JSB) has published detailed county statistics that are hardly matched by any other provinces (JSB 1989–97; 1994b). Besides an analysis of these data, I also synthesize information gathered from my numerous fieldwork and informal interviews with Chinese officials since the mid-1980s to understand the process of economic reforms and the forces underlying regional development.

Three major economic indicators have been selected to exhibit the trend of regional inequality.[5] The first indicator is national income (NI) (or net material product) per capita, which has often been used in the study of regional inequality in socialist countries (Lyons 1991).[6] The second indicator is gross value of industrial and agricultural output (GVIAO) per capita, which has also been employed as an effective indicator of economic development in China, particularly at the lower administrative level.[7] The third indicator is gross domestic product (GDP) per capita. Like many other socialist countries, China has traditionally used NI and GVIAO as major indicators of economic development, but GDP has recently been introduced. Data for comparable prices, calculated from growth indexes, are used to reveal temporal trends.[8]

Fiscal and investment systems play a significant role in socialist economies (Kornai 1980, 1992), including the Chinese economy (Wong 1991; Naughton 1995a). My analysis of investment focuses on fixed (asset) investment and foreign direct investment (FDI).[9] A fundamental feature of globalization is the increase of foreign investment in developing countries, including China. Foreign investment is exerting an increasing influence on China's regional development and should be analyzed. Regions or local conditions examined in this book include local states and factors that vary spatially, such as development paths, ownership structure, interregional linkages, and

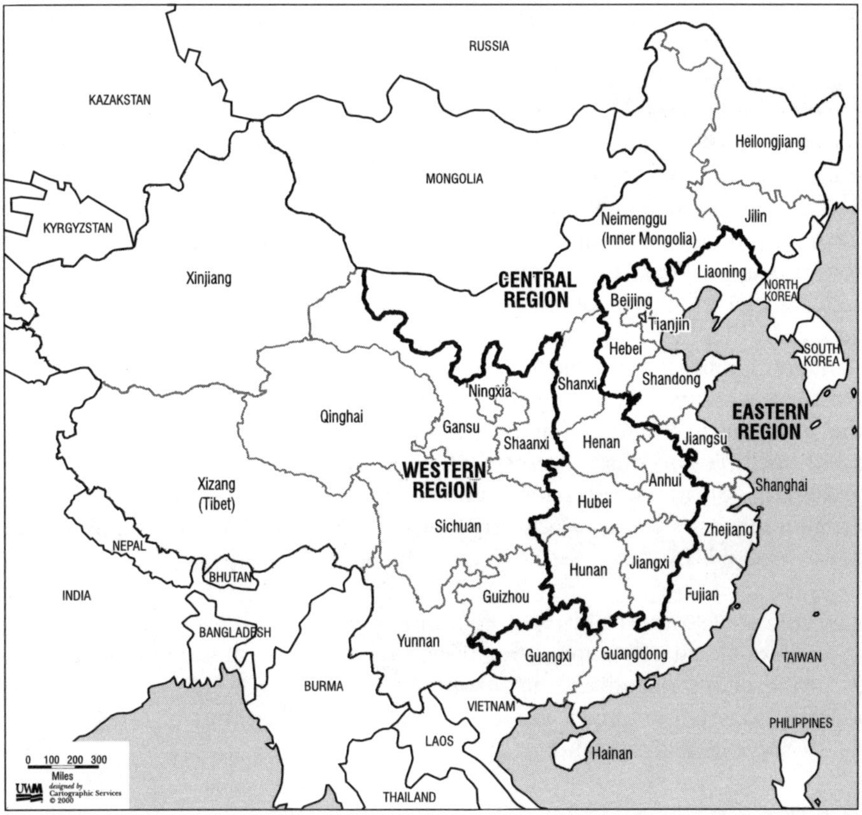

Figure 1.1 Regions and provinces in China.

resources. My analysis of state policies includes fiscal policy, investment policy, open door policy, and other policies that affect China's investment distribution and regional development. In terms of limitations, this research does not focus on labor migration, class relations, and production organization. Labor has been abundant in China, and during much of the last five decades, labor migration was controlled by the government, and thus considering the whole research period, labor mobility has exerted much less impact than capital on regional development.[10]

As I have argued in previous sections, regional inequality is sensitive to geographic scale, and multiscale seems to be an important feature of regional inequality in China. China is divided into eastern (or coastal), central, and western regions, based on the long-standing coastal–interior divide and following the practice adopted by the Seventh FYP (1986–90) (State Council 1986; SSB 1986) (Fig. 1.1). Another important dimension is inequality across provincial-level administrative divisions (provinces).[11] For brevity and clarity, regional inequality, in terms of the macro regions and provinces, will be referred to as interregional inequality and interprovincial inequality, respectively.[12]

Scholars have stressed the importance of analyzing the impact of economic reforms on economic, political, and spatial changes within the provinces (Goodman 1997), and the study of intraprovincial inequality can improve our understanding of regional inequality in China (Yang 1997; Wei and Fan 1997). As I will elaborate in Chapter 5, I have selected Jiangsu province for in-depth analysis of intraprovincial inequality in China. Jiangsu well mirrors the dramatic changes taking place in China and its growth trajectory reflects that of many other coastal provinces. Large regional disparities, especially between poorer northern Jiangsu (Subei) and richer southern Jiangsu (Sunan), have existed historically, and has generated renewed concerns of scholars and governments. Moreover, in Jiangsu one can observe the interplay and interaction of a multitude of forces of regional development, including foreign investment, state policy, and local initiatives. Availability of subprovincial data and my experiences with Jiangsu provide another incentive for my study. My fieldwork in Jiangsu has been particularly concerned with:

1 the process and policies of economic reforms,
2 central–local government relations (especially fiscal arrangements),
3 local policy initiatives and resources,
4 interregional linkages, especially ties to Shanghai, and
5 foreign investment and trade.

The coefficient of variation (CV), one of the most often-used indexes, is employed to measure regional inequality.[13] I use the location quotient (LQ) to reveal the departure of individual regions from the average.[14] I perform correlation and regression analyses to reveal the contributions of foreign investment, state policy, and local conditions to regional development, which are discussed in related chapters. Finally, I use geographic information systems (GIS), especially ARC/INFO and ARCVIEW, to develop databases, perform spatial analysis, and produce digital maps.[15]

1.6 Organization of the book

This book consists of seven chapters. After this introductory chapter, Chapter 2 first overviews China's state policy and regional development from 1949 to the mid-1990s. I demonstrate that Mao emphasized industrialization and national defense, while decentralization policy did provide some decision-making power to regions. I then examine changing patterns of regional inequality across China's macro regions and provinces. I show that regional inequality persisted under Mao, and during the reform period, interregional inequality rose, while interprovincial inequality declined. I further analyze the changing status of provinces, and reveal the rising status of Guangdong, Fujian, Zhejiang, Jiangsu, and Shandong, and the relative decline of traditional industrial bases.

Chapters 3 and 4 analyze and synthesize multiple mechanisms underlying regional development in China, mainly the central state, local states and agents, and foreign investment. In Chapter 3, I examine changing patterns of fiscal systems and fixed investment in great detail, and discuss the implications of these changes on regional development. Investment, for example, shifted to the Third Front area during the mid- and late 1960s, but the coastal region has taken the lead during the reform period. Fiscal and investment reforms have provided favorable conditions for a group of coastal provinces, and contributed significantly to their rapid growth during the reform period.

In Chapter 4, I examine changes in ownership structure and analyze regional patterns of foreign investment. I show how Mao's decentralization policy and local efforts contributed to the development of a non-state sector, which laid the foundation for its phenomenal growth during the reform period. I also demonstrate that the costal region has dominated foreign investment in China, and foreign capital is particularly important to those rapidly growing coastal provinces. I finally synthesize state policy, foreign investment, and local forces, for a better understanding of regional inequality in China.

Chapters 5 and 6 examine regional development within China's provinces through a case study of Jiangsu province. In Chapter 5, I examine changing patterns of regional inequality in Jiangsu. I reveal that regional inequality in Jiangsu persisted under Mao, contrasting to the normal notion that Mao reduced regional inequalities within provinces. During the reform period, interregional and rural intercounty inequalities increased, while overall intercounty inequality declined.

In Chapter 6, I explain these changing patterns by analyzing central government policy, local states and development conditions, and foreign investment. I show that during Mao's era, state policies in general were more favorable to the cities and Sunan, while local efforts were also important and the global context cannot be ignored. During the reform period, reform policies, local states and agents, and foreign investment have all contributed to the phenomenal growth of counties in Sunan, while Subei and the relative status of traditional cities have declined, producing complex landscapes of regional development in Jiangsu. The last chapter summarizes the major findings of this research and discusses their theoretical and policy implications.

2 Patterns of regional development

During Mao's era, China's economic systems were essentially centralized: the state attempted to control the Chinese economy, although state policy shifted constantly. Moreover, China was largely isolated from the outside world, as a result of Mao's self-reliance policy and the antagonism of Western countries towards China. While Mao's China, like other countries under state socialism, emphasized urban-centered industrialization through the development of heavy industry, national defense and spatial equality were also considered as major principles guiding regional development and industrial distribution. While considerable work has been done in understanding China's economic and political systems, scholars still disagree whether regional inequality under Mao declined, due to data constraints and differences in methods.

China since the late 1970s has implemented a series of economic reforms that have restructured Mao's development policies and China's regional economy. Although scholars have been debating over changing patterns of regional development in post-Mao China, the triple transitions of decentralization, marketization and globalization, as well as China's changing regional development policies, have dramatically changed the mechanisms of regional development. In this chapter, I first provide an overview of the process and policies of economic and regional development in China. This is followed by an analysis of the emergence of multiple mechanisms underlying regional development. I then examine the trends of interregional and interprovincial inequalities. I also investigate emerging patterns or forms of regional development through an analysis of the changing status of the provinces in the Chinese regional economy.

2.1 Development policy and regions under Mao

Before the establishment of the PRC in 1949, China's regional economy was spatially uneven, with the coastal region more developed than the interior. Indeed, distinct regions and regionalism have long existed in China, partly based on the varied physical geographical conditions, especially the three great topographic steps in China—the Tibetan plateau in the west; the central mountains and plateaus; and the eastern plains (Zhao 1994).

Chi (1936), mainly based on agricultural and transport conditions, divided traditional China into independent key economic areas. Skinner (1977) used a concept of macro regions for a better understanding of China's historical geography. Considerable disparity existed among the macro regions in which people spoke different languages and had different identities, forming the basis of regional disparities in China. These macro regions and their surrounding areas divided China into two parts: the more developed southeast (China proper), which consisted largely of Han Chinese and the more developed regions, and the northwest (extra-ecumenical China), which was largely occupied by minorities (Edmonds 1990). Moreover, China proper has most of the cultivable land and extra-ecumenical China is a mountainous area; the former receives more rainfall, enjoys more fertile soil, and has extensive rivers and coastlines, which provide more favorable conditions for agricultural, transportation, and industrial development (Pannell and Ma 1983).

While the Western countries were undergoing industrial revolution, China, after the Song dynasty (960–1279), entered into a period of stagnation and slow growth. The Opium War (1839–42) ended more than 2000 years of imperial rule in China and a semi-feudal, semi-colonial society emerged. The Treaty of Nanjing, which was signed in 1842, opened up five treaty ports (Shanghai, Guangzhou, Ningbo, Xiamen, and Fuzhou) for colonial control and penetration, followed by the opening up of more treaty ports and the cession of territory (Murphey 1974). Early in the 20th century the northeast region (Manchuria) was controlled by the Japanese who quickly established heavy industrial bases there to support their war efforts. While introducing modern technology, Western imperialism exerted economic, political, and military control over China and ruined China's nascent native industries and rural economies. Although treaty ports, particularly Shanghai, enjoyed economic prosperity, the vast majority of China was in poverty. The Chinese economy suffered severe damage during World War II and the Civil War (1946–49).

When the PRC was established in 1949, China was troubled by chaos, severe depression, high inflation, and huge income and regional inequalities. The coastal region produced about 70 percent of industrial products while the vast interior was devoid of modern industries, except a few cities such as Wuhan and Chongqing. Shanghai, Tianjin, Jiangsu, Guangdong, and Manchuria were considerably more developed than the rest of the nation. Much of the railway and port development was also concentrated in coastal China and Manchuria (Wu 1967). Mao's new socialist regime faced a poor economy and a colonial legacy of regional imbalance.[1]

During the Period of Recovery and the First Five-Year Plan (FYP) periods (Table 2.1),[2] China launched the Soviet-model of industrialization and the socialist transformation of private enterprises, and adopted the collectivization of agriculture in the countryside.[3] With the help of Soviet experts, China established a highly centralized mode of planning and administration to take control of the economy and to finance industrialization, including the

Table 2.1 Major periods of national development in China

1949–1952	Period of Recovery	Founding of the PRC in 1949
1953–1957	First Five-Year Plan	
1958–1962	Second Five-Year Plan	Great Leap Forward (1958–60)
1963–1965	Period of Readjustment	Third Front Construction (1964–71)
1966–1970	Third Five-Year Plan	Cultural Revolution (1966–76)
1971–1975	Fourth Five-Year Plan	
1976–1980	Fifth Five-Year Plan	Launch of Economic Reforms in 1978
1981–1985	Sixth Five-Year Plan	
1986–1990	Seventh Five-Year Plan	Tiananmen Incident in 1989
1991–1995	Eighth Five-Year Plan	
1996–2000	Ninth Five-Year Plan	

establishment of the State Planning Commission and the State Construction Commission. The First FYP period emphasized rapid industrial growth through the development of urban-centered heavy industry. The state gradually transformed private enterprises into state-owned enterprises (SOEs) operating under planning.

Regional development has since become an important component of policies in China, due largely to its role in the debates over the nature of socialism, central control and local autonomy, and resource allocation among regions (Bo 1993). Influenced by socialist ideology and egalitarian ideas, Mao attempted to develop backward inland areas and to reduce regional imbalances. First, the leaders stressed rapid industrialization through the development of heavy industry. Following the Soviet principles of industrial allocation, the leaders believed that industrial production should be close to energy and raw materials.[4] Since much of the industrial base was in the coastal region and raw materials in the interior region, efforts were taken to build industry in the interior. Secondly, as the perceived military attack on China would come from Taiwan, Korea and other places near the coast, major efforts were made to establish industries in the interior for defense considerations. Finally, the new leaders considered inequality a colonial legacy that should be eradicated with the establishment of socialism, as one objective of socialism was to eliminate income and spatial inequalities. Balanced regional development was one of the major factors considered in regional allocation of resources during the 1950s.

The relationship between the coastal and interior regions was most clearly expressed in Mao's highly influential speech 'On the Ten Great Relationships.' Mao stated that

> 'in the past our industry was concentrated in the coastal regions .... This irrational situation is a product of history. The coastal industrial base must be put to full use, but to even out the distribution of industry as it develops we must strive to promote industry in the interior .... Without doubt, the greater part of the new industry should be located in the

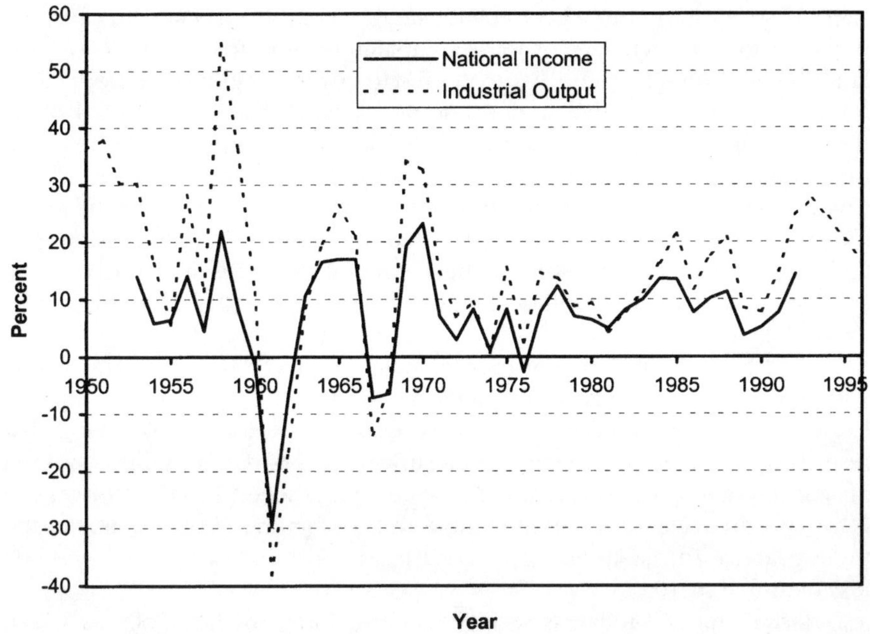

Figure 2.1 Annual growth rates of national income and gross value of industrial output, 1950–95.
Source: SSB 1990–97.

> interior so that industry may gradually become evenly distributed; moreover, this will help our preparations against war . . .'
>
> (Mao 1956, pp. 286–7)

Such concerns for national defense, spatial equality, and economic efficiency were identified as the primary goals of the First FYP (State Council 1956; Bo 1993). After three years of recovery (1949–52), massive construction was soon under way. The initiation of the First FYP was centered on the construction of the 156 industrial and energy-based projects, all large in scale, capital intensive, and imported from the former Soviet Union. About a half of these projects were allocated in central China, but the state never invested heavily in the neighboring Taiwanese provinces of Fujian and Zhejiang.

China experienced rapid industrial growth and steady agricultural growth during the First FYP period, although the living standards in China remained low (Fig. 2.1). With the initial success of socialist transition, Mao called for faster economic growth and radical social change by launching the Great Leap Forward (GLF). Completely unrealistic targets were set for industrial and agricultural growth, especially for the iron and steel industry and for rice production. Rural people's communes were established as a higher stage of socialist organization to eliminate the means for farmers to carry out private economic activities. A mass campaign of industrial construction was launched and rural peasants were ordered to engage in industrial production. Meanwhile, administrative and economic decentralization relegated some decision-making powers to localities. However, socio-political unrest and unrealistic means of industrialization soon led to the collapse of the Chinese economy. The GLF, intensified by inclement weather and the termination of Sino–Soviet relations, resulted in social chaos, recession and the great famine of 1959–61. At the height of the crisis in 1961, national income and industrial output declined 29.7 percent and 38.2 percent, respectively (Fig. 2.1).[5]

With the severance of Sino–Soviet relations in 1960, the Soviet Union withdrew all its technicians and aids. In the years that followed, Chinese leaders abided by self-reliance and focused on economic adjustment that resembled the guiding slogan: 'readjustment, consolidation, filling out, and raising standards.' Consequently, the state strengthened its control over the economy, and closed or curtailed thousands of factories.[6] But even after the three years of readjustment, per capita food consumption in 1965 was still less than that in 1957.

Defense and self-reliance received high priority in the 1960s as China adjusted to the deterioration of Sino–Soviet relations and the intensification of the Cold War. China's perception of threats of foreign attacks represented a long-term concern for national security that began with the Korean war in the early 1950s and the severe bombing of Taiwan-controlled islands of Jinmen and Mazu in the late 1950s. With intensified Sino–Soviet ideological and territorial disputes and increasing American military involvement in Vietnam, China's perception of international threats grew rapidly. Mao and the top leadership declared in May 1964 that China should prepare for wars and strengthen the construction of the Third Front (*san xian*) areas. As I will elaborate in Chapter 3, the Third Front policy tied up much of state investment and resulted in the construction of numerous defense-oriented industries in the interior provinces (Naughton 1988; Ma and Wei 1997).

The Third Front construction was followed by another damaging event—the launch of the Cultural Revolution in 1966. Mao called for class struggle and mass mobilization to throw out 'capitalist roaders,' authority and 'olds' under such slogans as 'class struggle is the driving force of social development' (Riskin 1987). Private commerce and markets were prohibited as components of capitalism, and fewer goods were available for consumers. Numerous government organizations, factories, and educational institutions were disbanded. During high tide (1966–68), the movement was particularly disruptive and violent. Consequently, national income and industrial production dropped sharply (Fig. 2.1), and the living standards of the people stagnated.

China experienced some policy changes during the early 1970s. The death of Lin Biao in 1971, who was instrumental to the Third Front program, marked the termination of this program and the emergence of a pragmatic leadership. With its entrance into the United Nations in 1971, and the normalization of Sino–US relations, China's isolation from the West abated. With the change of leadership in the Soviet Union, the perceived threats of wars vanished, and national defense gradually lost its priority. In addition, problems with the Cultural Revolution and Third Front construction forced the leadership in China to adjust its economic and regional policies. Intense efforts ensued to develop China's petroleum resources in the coastal provinces, especially in Shandong, Tianjin and Hebei. Moreover, substantial investment was earmarked for the improvement of port facilities along the coast and in the lower reaches of the Yangtze River. From 1971 to 1976, average growth rate of national income remained low (4.1 percent), while shortage and slow growth failed to raise the living standards of the people.

China's economic and development problems under Mao included:

1 overly centralized planning systems, poor planning, and economic mismanagement;
2 overemphasis on national defense and the lack of efficiency;
3 overemphasis on self-reliance;
4 excessive emphasis on grain production and the low productivity of commune systems;
5 excessive accumulation of national income at the expense of consumption;
6 overemphasis on heavy industry while ignoring light industry and services (Riskin 1987; Hsu 1991).[7]

Class struggle, mass movement, and political unrest, especially during the GLF and the Cultural Revolution, seriously damaged the Chinese economy. Regional development policy during Mao's era was also criticized for ill-guided industrial allocation principles and poor investment efficiency (Li *et al.* 1990; Lu *et al.* 1990). These problems, as well as the changing domestic and international conditions, made it necessary for the reform of China's economic systems.

2.2 Post-Mao reforms and regional policy

After the death of Mao in 1976 and a short transition period under Hua Guofeng, a new pragmatic leadership under Deng Xiaoping emerged. China formally launched economic reforms in 1978, and followed with a series of reform policies in the next two decades. China has also adopted a flexible attitude toward socialism and capitalism. Orthodox socialism is no longer seen as appropriate for China, and 'socialism with Chinese characteristics' has been advocated as the ideology for the nation. The government sees China as a country at the initial stage of socialist development, during which facilitating economic growth and raising the living standards of the people are considered as the dominant development objectives, rather than 'class struggle'. Economic reforms have dramatically restructured China's development policy and regional economies.

Announced in December 1978, rural reforms decollectivized the rural communes, and introduced the rural responsibility system. Farmers received decision-making powers on their economic activities, although they have to fulfill the contracted quotas of grain production required by the government. Relative slow growth of urban production necessitated urban enterprise reforms. Launched in October 1984, urban reforms allowed market forces to adjust the distribution of commodities and materials, and encouraged the development of urban services. After a short period of transition (1989–1990), China, in the early 1990s, launched far deeper economic reforms, and advocated development of a socialist market economy.

Meanwhile, China confronted Mao's policy of self-reliance by favoring the opening up of its domestic economy to the outside world. To encourage foreign investment, in 1979, China introduced the Joint Venture Law which permitted foreign investment in China, followed by the establishment of four special economic zones (SEZs), which received various policy incentives. China's open areas were extended in order from SEZs to 14 open coastal cities (OCCs) in 1984, to three delta areas (Zhujiang, Minjiang, and Changjiang deltas) in 1985 and 1986, and later to Hainan Island in 1988. A series of laws and incentive policies were introduced in the mid- and late 1980s to encourage foreign investment and trade. The government gradually improved relations with foreign investors, especially investors from Taiwan, South Korea, and Singapore. By 1997, about a half of Chinese cities and counties were opened up to foreign investment and trade.

Two decades of reform have brought dramatic growth in China, which can hardly be matched by any other country (Fig. 2.1). Average annual growth rate of per capita gross domestic product (GDP) was 7.8 percent during the 1980s. The growth was accelerated in the early and mid-1990s, as average annual growth rate of per capita GDP reached 10.3 percent during the 1990–96 period (Table 2.2). The living conditions of rural peasants have also been improved greatly, although the gap between the city and the countryside has re-surged in the 1990s after an initial decline during the early 1980s. However, the pace and consequence of economic reforms are uneven across regions and cities (e.g., Xie and Costa 1991; Goodman and Segal 1994). As I will show in following sections, the coastal region has, in general, recorded more rapid growth than the interior region. Coastal provinces of Guangdong, Jiangsu, Zhejiang, Fujian, and Shandong have experienced the fastest growth among China's provinces.

China's regional policy has also experienced dramatic changes. Regional policy under Mao was criticized for its emphasis on regional balance, self-reliance and low efficiency (Xue 1982; Li *et al.* 1990). With the shifting of China's dominant development policies from self-reliance to open door policy and comparative advantage, China's reforms in the 1980s provided

Table 2.2 Major economic indicators in China, 1980–96

	1980	*1990*	*1996*	*1980–96 average annual growth (percent)*	*1990–96 average annual growth (percent)*
Population (million)	987	1,143	1,224	1.4	1.1
GDP (billion yuan)	452	1,855	6,859	10.1	11.6
Per capita GDP (yuan)	460	1,634	5,634	8.7	10.3
Fixed assets investment (billion yuan)	91	452	230	21.9	34.6
Revenue (billion yuan)	116	294	741	12.3	16.7
Foreign investment (US$ billion)	—	10.3	54.8	—	32.1
Export (US$ billion)	18.1	62.1	151.1	14.2	16.0

Source: SSB 1997.

favorable policies to some coastal areas. The 'ladder-step theory' (*tidu lilun*), a Chinese version of the growth pole and inverted-U theories, was proposed. According to this theory, China, as a developing country, should concentrate its resources in the more developed coastal region, which is akin to a higher step; gradually the emphasis of development will be shifted to the central and western regions, and the diffusion of the coastal development will stimulate the development of the interior. Influenced by the ladder-step theory, the government encouraged regions with comparative advantages in factor endowments, largely located in the more developed coastal region, to get rich first. Cities and core regions are considered growth poles or engines of growth and are emphasized by government policy, although China's urban policy still attempts to control the size of large cities. The Sixth FYP (1981–85) encouraged regions to develop their economies based on comparative advantages and spatial division of labor.

A 'three economic regions' strategy was introduced in 1985, in which the coastal region was expected to develop high-tech industries and to actively participate in the international market, while the central and western regions were earmarked for energy, agriculture, and mineral development. The government hopes that the rapid growth of the coastal region will eventually function as a powerful engine to propel national and interior development. These ideas of regional development were explicitly stated in the Seventh FYP (1986–90) (State Council 1986). It emphasized efficiency-oriented coastal development, which is expected to eventually help the development of the interior through diffusion and cooperation. A series of reform policies in investment, finance, and trade were implemented and provided more decentralized and more open policies for the coastal region, especially for special economic zones, coastal open cities/zones, and numerous economic and technological development districts.

The coastal development strategy (CDS) was officially decreed in 1988 and was designed to facilitate China's open door process and to enable China to compete in the global market through the development of the coastal region. CDS directed that the coastal region should participate more in international markets and develop export-oriented economies. Industrial production in the coastal region should be 'putting both ends (sourcing and marketing) towards the global market.' The coastal region received favorable policies through the implementation of the CDS, and its foreign investment and trade increased rapidly in 1988. In summary, during the 1980s, major policies favored coastal development, including:

1 more state investment was allocated to the coast;
2 the coast enjoyed more tax breaks for export, higher foreign exchange retention rates and lower tariff in import;
3 decentralization gave the coastal region more autonomy; and
4 policies favoring the coast also attracted labor and raw materials from the interior.

But some special policies were also granted for the development of the backward regions, such as fiscal subsidies and the establishment of border-region open areas.

After a short period of stagnation following the Tiananmen incident in 1989, the 1990s witnessed the implementation of bolder reforms towards developing China into a socialist market economy. In January 1992, Deng Xiaoping toured southern China and pressed for further economic reforms. This was followed by the Fourteenth Party Congress of October 1992 calling for deeper reforms and the rapid development of market economies. Meanwhile, preferential policies had been granted to Shanghai to revitalize its stagnant economy. Several categories of development zones have flourished in China, especially in the coastal region, including economic and technological development zones, free-trade zones, export-processing zones, and border trading zones. Even small towns in coastal China have established numerous local-level economic and technological development districts. During the early and mid-1990s, economic growth in China was unprecedented, as evidenced by the double-digit growth from 1992 to 1996. Some regions in China, particularly some coastal areas such as southern Jiangsu, recorded dramatic growth.

However, issues of inflation, central control and local autonomy, the rich and the poor, and regional inequality remain major points of debate. Scholars have warned against the rise of income and regional inequalities that is threatening economic growth and social stability. Millions of rural workers migrated to coastal cities for jobs, generating considerable concerns over the impacts and policies of migration and urbanization in China. Regional 'resource wars', in which localities using illegal administrative and military measures to protect their local markets and restrict interregional resource flows, broke out in many regions (Wedeman 1993). The bargaining of the central and local governments over resource allocation and development policies became extremely intense, as China's poor areas believed that they were less favored by the central government (Wong 1991; Wei 1996). These problems, intensified by inflation and corruption, prompt many to consider regional inequality to be a root of China's regional problems and an important issue of government policies (Fan 1997; Yang 1997). The ladder-step theory was criticized for being too crude as a foundation for China's regional policy, and the implementation of such a policy intensified regional gaps and regional conflicts, rather than prosperity for all. In addition, globalization and the technology revolution may provide opportunities for the faster growth of poorer regions. The interior provinces wanted a fair treatment by the central government, and demanded more resources for interior development.

With the intensification of coastal–interior relations and the critique of uneven development strategy, the government reduced policy emphasis on the coastal provinces and encouraged interregional cooperation and the development of the Yangtze River Valley. The Ninth FYP (1996–2000) perceives polarization as a serious threat to China's prosperity, stability, and unity, and places reducing regional inequality as a top policy priority. The government promised to commit more resources to reduce the incidence of poverty and to develop the interior region. More favorable policies have been extended to the interior region, especially in investment allocation, fiscal reforms and open door policies. The 1994 fiscal reform improved central control of fiscal resources, which has strengthened state capacity for funding interior development. Indeed, the share of the interior region in key-point project investment has risen since the mid-1990s, and more efforts have been made in poverty relief and the development of the poorest areas. The government has been seeking more participation by international organizations in poverty relief and interior development. It has encouraged economic cooperation between the coastal and interior regions, and the transfer of labor-intensive industries from the coast to the interior. In 1997, Chongqing was designated as the fourth centrally administrated municipality of China, in the hope of developing Chongqing into a growth pole for western China. The government has changed some policies towards foreign investment and trade to reduce the bias towards foreign investors and the coastal region.

2.3 Changing mechanisms of regional development

Economic reforms have fundamentally changed the mechanisms underlying economic and regional development in China. A burgeoning literature has attempted to explore the nature of economic transition in China. One group of scholars has emphasized the decentralization of power from the central state to local states and considers that local states have the leading role in economic development. Oi (1992), for example, conceptualizes China's economic transition as a process of decentralization, during which local states have taken control of economic development, giving rise to local state corporatism. Another group of scholars highlights the transition from state socialism to market socialism and the notion of path dependence. For example, Nee's (1989) market transition theory emphasizes the infusion of market mechanisms as fundamental to the change of income distribution in rural China. These two perspectives, however, have limitations when applied to the understanding of the change of the Chinese regional economy, and have only partially captured the nature of China's economic transition from state socialism. I suggest that China's economic transition can be best understood as triple processes of decentralization, marketization and globalization, rather than decentralization or marketization. An analysis of this triple transition is helpful for a better understanding of the changing mechanisms underlying regional development in China.

In orthodox socialist countries, the state attempts to monopolize the control of economic activities, largely through centralized fiscal, investment, and planning systems. Local states are often assigned the task as agents of the central state. Shortly after the founding of the PRC in 1949, China formulated a centralized system of finance, investment, and administration to take control over the economy and to launch the Soviet model of industrialization. While Mao's system was essentially centralized, China decentralized its economic systems during certain periods, especially during the late 1950s and the early 1970s. During these periods, administrative authority of enterprises was partly transferred to local governments, and numbers of centrally controlled enterprises and products were slashed (Riskin 1987). Regions were encouraged to develop small-scale enterprises based on local resources and markets, and were given the main responsibilities for capital construction, resource distribution, budgeting, and enterprise management. Decentralization provided some incentives for local governments to stimulate economic growth, and local governments and enterprises also obtained management experiences from decentralization. In other words, before economic reforms, regions differed greatly in their experiences in managing local economies and in their abilities to adapt to changing macro conditions. However, decentralization often resulted in economic chaos and mismanagement, largely due to problems in planning and political struggle.

Post-Mao economic reforms have brought dramatic changes to the level of central control and the role of states. Economic and administrative decentralization to local governments and enterprises is an important theme of reforms. Decentralization attempted to reduce the level of central control over decision making and to provide more incentives to regions. As a result, reforms have given local authorities considerable power in decision making and resource allocation. Although the state remains important in China, the control and capacity of the central state have declined, while local governments have become more important than before in economic development.

Fiscal reforms have reduced the fiscal control of the central government, and provided local governments considerable powers in revenue collection and local spending. The People's Bank of China no longer monopolizes banking functions, and local governments are heavily involved in the operation of local financial institutions. Investment reforms have increased the control of local governments over capital resources. In addition, decentralization has provided local governments considerable power in enterprise management. These changes will be further demonstrated in the analysis of fiscal systems, investment patterns, and industrial ownership structure in the following chapters.

Decentralization has allowed local governments to actively initiate policies to stimulate local economic growth, and makes a great difference to reform policies that differ from region to region. Local governments, especially township and village governments, are acting very much as entrepreneurs whose motives derive from a desire to maximize income and power. Township and village officials are often the managers and technical know-how of rural enterprises, and are directly involved in the production process. The interdependence between local governments and enterprises stimulates the bargaining effort of local governments (Walder 1995b). During the reform period, local governments have actively negotiated with the central state for more favorable policies. However, the tension of central control and local autonomy has persisted in China, and waves of decentralization tend to be followed by attempts of recentralization by the central government. Moreover, decentralization alone is inadequate to characterize the process of China's economic reforms. Marketization is another major theme of reforms and has become especially important during the 1990s.

The relaxed state control and the growth of market mechanisms in economic and spatial development can be demonstrated in many aspects. First, the control of state budget over the economy has declined, as evidenced by the decrease of the ratio of budget revenue to GDP from 25.7 percent in 1980 to 15.8 percent in 1990 and 11 percent in 1996. The control over local expenditures has also been reduced; enterprises and individuals have obtained considerable fiscal authority. Secondly, state control over investment has steadily declined while proportions and absolute sizes of investment controlled by localities have increased. Budgetary investment, as a percentage of total investment, declined dramatically from 77.7 percent in 1978 to only 6.6 percent in 1995. State budgetary investment, as a source of financing for fixed asset investment, declined from 28.1 percent in 1981 to 3.1 percent in 1995. In 1995, enterprise funds and domestic credits provided 53.2 percent and 21 percent of fixed asset investment, respectively, followed by foreign investment (11.5 percent). Thirdly, the price reform has dramatically reduced state control over prices. A two-track price system (market price vs. state fixed price) was created in the early and mid-1980s, but market prices have eventually taken control. From the late 1970s to 1994, the share of industrial output subject to the plan fell from 95 percent to 4 percent

(Chen *et al.* 1995). By the late 1990s, virtually all retail goods were sold at market prices, rather than state-fixed prices. Lastly, shares of output and labor absorption by SOEs have declined dramatically. The share of output produced by SOEs declined from 77.6 percent in 1978 to 34 percent in 1995. In many coastal provinces, the state sector has become a minor sector. For example, in Zhejiang, SOEs only produced 14 percent of GVIO in 1994. Multiple forms of ownership have emerged in China, including state, collective, private/individual, foreign, shareholding, and joint ownership forms. Non-state sectors have grown rapidly and are exerting a great influence on the Chinese economy.

The reform of China's planning system is critical to the changing role of planning and markets. In the early and mid-1980s, Soviet-style mandatory planning was gradually discarded in favor of guidance planning to establish a planned socialist commodity economy. A two-track price system was created to introduce the market mechanism gradually. In 1987, the policy was coined as 'the state regulates the market, and the market guides the enterprises.' This policy was changed in 1992 in order to develop a socialist market economy. Consequently, many bureaucratic coordination organs and vertical control instruments have been removed. The state has become more flexible, entrepreneurial, legalistic, and technocratic than its predecessor (Howell 1991). The state has been changing from planning and directing economic resources directly to managing the policies, which adjust the resources and markets (World Bank 1992a). It is exerting much less control over the economy than in Mao's era, and is no longer monopolizing economic development in China. Indeed, some scholars were very concerned that the state had lost its control over the Chinese economy (Wang and Hu 1993). But the process of China's economic reforms is not simply removing state control (Wedeman 1999); the state still actively initiates macro policies that are critical to economic development. In addition, the processes of decentralization and marketization are accompanied by another process—the globalization of economic activities.

China has implemented a series of 'open door' and reform policies to attract foreign investment, reversing Mao's policy of self-reliance. China's opening up to the outside world has been facilitated by the globalization of economic activities, during which foreign investment and trade have increased on the global scale. Indeed, a large amount of foreign capital has been invested in the newly industrializing countries (NICs) of Asia, and recently, in China. A number of structural, technological and institutional forces have facilitated the globalization of economic activities (Malecki 1991; Dunning 1993). Globalization increases partly due to low profitability and intense competition in developed countries, and to cheap labor in developing countries. Developments in transportation and communication and changes in technology and production organization have also facilitated the relocation of firms overseas. Meanwhile, foreign investment and trade have been widely viewed by developing countries as a way of achieving economic growth and benefiting from globalization and the new international division of labor. Governments in these countries have promoted opening up to the outside world and globalizing their domestic economies. These changes, as well as strategic alliances and diminishing barriers, have all contributed to the trend of globalization, the increase of FDI, and the emergence of new geographical spaces. Without these global changes and the participation of overseas Chinese, China's policies for attracting foreign investment would be much less effective.

China's potential market, cheap labor and policy incentives have lured investment from both multinational enterprises (MNEs) and small investors. From 1991 to 1995, FDI increased dramatically from $4.4 billion to $37.7 billion. Rapid economic growth, foreign investment, efforts of local governments, and ties to the international markets have also contributed to the growth of international trade. Exports increased from $9.8 billion in 1978 to $148.8 billion in 1995, and in 1995 imports reached $132 billion. In much of the reform period, foreign investment and trade were less significant. Foreign investment as a source of financing accounted for 3.6 percent in 1985 and 6.25 percent in 1990. However, by the mid-1990s, foreign investment and trade had become important components of the Chinese economy. In 1995, FDI in China accounted for 11.5 percent of fixed investment, and exports accounted for 21.5 percent of GDP although foreign investment and trade are unevenly distributed among the provinces. China has also opened its door to international financial markets and quickly developed its external bank networks. By the end of 1995, there were more than 100 foreign bank branches operating in China. In other words, globalization has emerged as one of the major forces shaping economic and regional development in China. Although some restriction policies were introduced in the mid- and late 1990s, China is much more integrated into the global economy.

In summary, China's economic reform is a pragmatic and gradual process, and can be understood as triple transitions of decentralization, marketization, and globalization. Decentralization or marketization alone only partly account for the process of China's economic reforms. These triple transitions have fundamentally changed major agents of economic and regional development in China. While the role of the central state remains important, local and global forces have emerged as equally important forces in driving regional development in China. Consequently, the central state, local states and agents, and foreign investors have become the three sets of dominating forces driving China's regional development.

2.4 Trends of regional inequality

As I mentioned in previous sections, development policy under Mao was very different from post-Mao reform policies. During Mao's era, considerations for economic growth and national defense were given high priority, while regional balanced development was also advocated. While economic and regional policies changed frequently, some effort was made to develop the interior, especially during the 1950s. However, post-Mao reforms have emphasized comparative advantages and an open door policy. Regional policy tends to favor more developed coastal regions, although interior development is not ignored and the government has re-stressed the importance of interior development. Patterns of regional development, especially during Mao's era, to a certain extent, reflect these policy changes. But during the reform period, as I will analyze in the next two chapters, local states and global forces have also greatly influenced the changing economic landscape in China.

2.4.1 Regional inequality under Mao

During Mao's era, some efforts were made in China to develop the backward interior region, particularly during the 1950s. But these efforts were interrupted by the defense policy and political unrest during the 1960s and 1970s. As evidenced in Fig. 2.2, both interregional and interprovincial inequalities increased, as shown by the change in the coefficient of variations (CVs).[8] Trends of interprovincial inequality were broadly similar to interregional inequality, but fluctuated more strongly.[9] Overall, regional inequality persisted during much of the 1950s and 1960s, but rose rapidly during the late 1960s and 1970s.

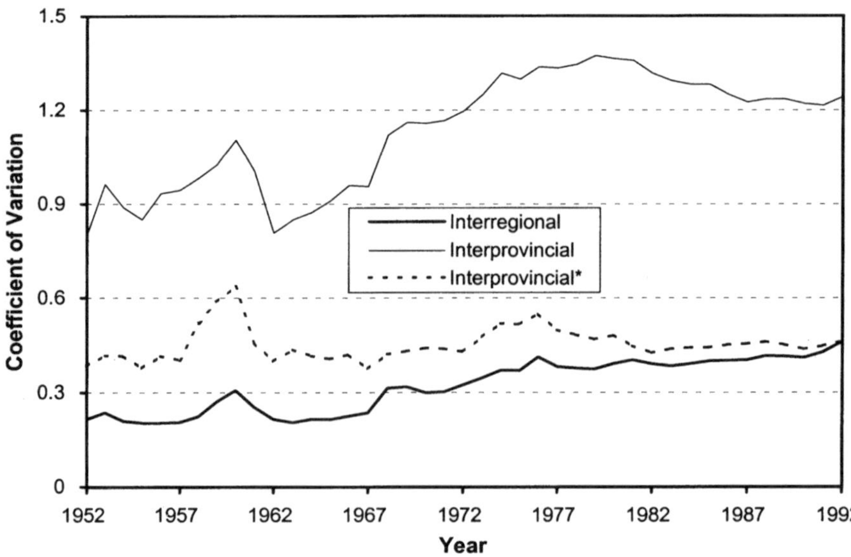

Figure 2.2 Trends of regional inequalities in China based on per capita national income, 1952–92.

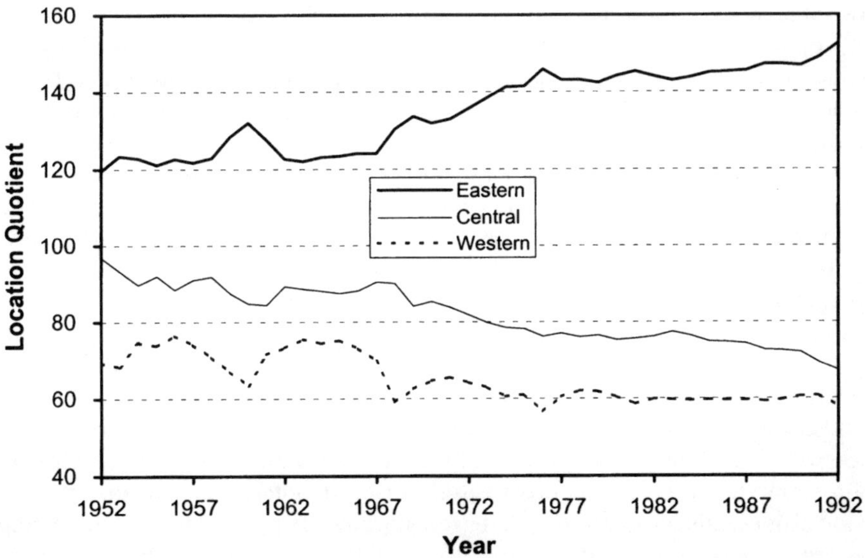

Figure 2.3 Changing status of regions based on location quotients of per capita national income, 1952–92.

When the PRC was established, the coastal region had developed much ahead of the interior. In 1952, the coastal region, with 14 percent of the total area and 42.3 percent of the population, produced 68.1 percent of gross value of industrial output. During the First FYP period (1953–57), an emphasis of new construction was placed on interior development for the consideration of industrialization, national defense, and spatial equality, while the recovery of the old industrial bases in the coastal region was also emphasized. These efforts facilitated the development of western regions somewhat, as evidenced by the rise of location quotients (LQs) for the western region (Fig. 2.3). Thus interregional and interprovincial inequalities did not increase further during the early and mid-1950s.

However, regional inequality rose greatly during the Great Leap Forward (GLF) (1958–60), when decentralization placed considerable decision-making power in the hands of regions, particularly the coastal provinces. But economic recession and famine following the failure of the GLF forced the Chinese government in the early 1960s to shut down thousands of construction projects and to disperse the urban population to the countryside. The economies of the three municipalities, which were always at the center of political and social movement and had the highest per capita income, went into deep crisis, while the poorest provinces were less affected politically and economically. With the decline of the leading provinces, both interregional and interprovincial inequalities decreased during the early 1960s.

From the mid-1960s to the early 1970s, China experienced a massive defense build-up in the interior, and in 1966 the tumultuous Cultural Revolution was initiated. Although much of the defense spending was in China's interior region, regional inequality fluctuated greatly but did not change much. As I will elaborate in the next chapter, this to a certain extent reflects the inefficiency of defense-oriented interior investment. However, on the eve of the initiation of another round of decentralization in the late 1960s, inequality increased, indicating that decentralization contributed to the improving conditions of richer provinces and the rise of regional inequality.

With the death of Lin Biao in 1971, the most powerful person after Mao, a pragmatic leadership gradually emerged in the early 1970s. China normalized its relationships with the United States and gradually abandoned the Third Front program. China's development policy shifted away slightly from the interior to favor the coastal region. In many coastal provinces, such as Jiangsu and Zhejiang, urban industrial production was improved and numerous rural township and village enterprises (TVEs) were established. As Figs. 2.2 and 2.3 show, from the late 1960s to the mid-1970s interregional and interprovincial inequalities in China increased rapidly; in particular, CVs revealing interprovincial inequality rose from 0.95 in 1967 to 1.35 in 1978.

Ironically, while Mao was calling for balanced regional development, Mao's period as a whole recorded a rise in regional inequality. From 1952 to 1978, the CVs for the regions and provinces increased from 0.22 and 0.8 to 0.38 and 1.35, respectively.[10] During the same period, LQs of the coastal region increased from 119 to 143, while those in both the central and western regions declined (see Fig. 2.3). While regional inequality persisted from the 1950s to the mid-1960s, inequality between the coast and interior regions was intensified after the late 1960s, and income gaps among China's provinces also widened.

The gap of per capita national income between the richest and the poorest provinces, Shanghai and Guizhou, increased in the 1952–78 period as well. In 1978, Shanghai, Beijing, Tianjin, Liaoning and Heilongjiang registered the highest national income per capita, followed by Jiangsu, Qinghai, Jilin, Guangdong, and Hebei, while the southwestern provinces remained the poorest in the nation. It is clear that after three decades of socialism, the basic patterns of regional development in China did not change much, and during Mao's era, regional inequality even increased somewhat or at least persisted. These findings greatly challenge the notion that Mao reduced regional inequality. I will later explain the persistence of regional inequality under Mao that contrasts sharply with the socialist ideology of regional equality.

I have also calculated trends of regional inequality by excluding the three municipalities, as shown in Fig. 2.2 (interprovincial* inequality). While broad patterns of change among interregional, interprovincial, and interprovincial* inequalities are similar—all rose under Mao, the extent of rising inequality is slightly smaller in terms of interprovincial* inequality.

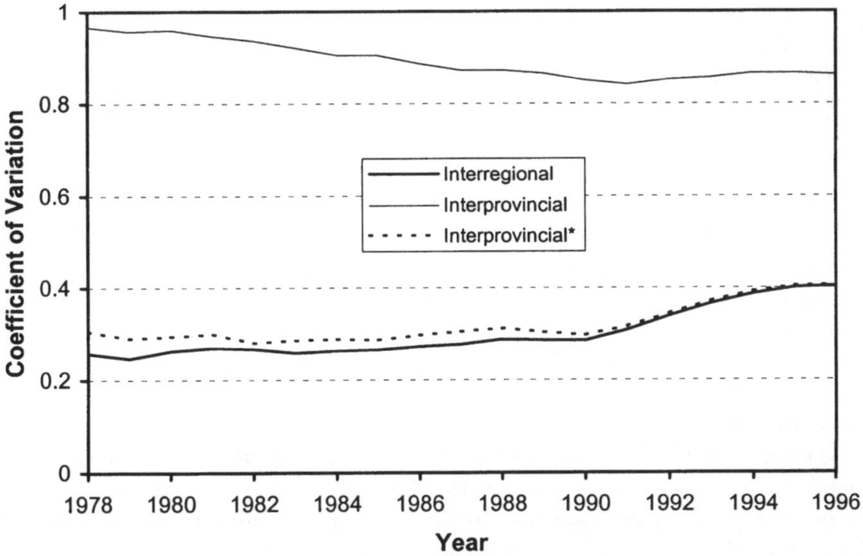

Figure 2.4 Trends of regional inequality based on per capita GDP, 1978–96.

This finding demonstrates that the three municipalities played an important role in the rise of interregional and interprovincial inequalities under Mao.[11] This also indicates that the intensification of urban–rural inequality, largely related to Mao's industrialization policy favoring cities, to a certain extent contributed to the persistence and rise of regional inequality in Mao's China.

2.4.2 Regional inequality in the reform period

During the post-Mao period, interregional inequality increased further, as evidenced by Figs 2.2–2.4. Moreover, LQs based on per capita GDP for the eastern region increased from 129 in 1978 to 145 in 1996, while that for the western region declined from 71 to 58. The rate of increase was relatively faster in 1984–88 and 1991–95 when bolder economic reforms were implemented. China's reforms during the earlier years were centered on agricultural reforms and the countryside, which tend to be less beneficial to rich urban regions than other reforms. However, the initiation of urban and enterprise reforms in the mid-1980s greatly benefited urban regions, especially more developed coastal regions. In addition, during the mid-1980s China opened up 14 coastal cities for foreign investment and trade. In some coastal provinces, such as Guangdong and Jiangsu, economic growth was accelerated by these policies. In 1988, more preferential policies were granted to China's coastal region with the announcement of the coastal development strategy and a series of reforms that favored the coastal region. Although it was interrupted due to the Tiananmen incident in 1989, regional inequality did rise in 1988.

The extent of China's economic reforms has been unprecedented from 1991 to the mid-1990s. Figure 2.4 shows that the most rapid increase in interregional inequality in the post-Mao period was recorded in 1991–95. CVs based on per capita GDP increased from 0.32 in 1990 to 0.44 in 1995.[12] During this period, China recovered from economic stagnation after the 1989 Tiananmen incident, and deepened economic reforms. Preferential policies were granted to Shanghai's Pudong district, and Shanghai has been encouraged to build into the engine of economic growth for China, particularly for the Yangtze River Valley. The coastal provinces aggressively restructured their economies and were further opened up to foreign investment and trade. The eastern region outgrew the central and western regions by a great margin.

It is clear that each time when deeper reform policies were implemented, regions, the coastal region in particular, received more decision-making power, retained more resources for local economic growth, initiated bolder local policies, and attracted more foreign investment. Favorable state policy, local initiatives, and foreign investment stimulated the more rapid growth of the coastal region, and led to the rise of interregional inequality during the reform period. But the trend of rising coastal–interior inequality lessened in 1996, indicating that interior growth had been accelerated.

Interprovincial inequality, however, declined, as shown in Figs 2.2 and 2.4. From 1978 to 1992, interprovincial CVs based on per capita national income declined from 1.35 to 1.24. CVs based on per capita GDP declined from 0.97 in 1978 to 0.84 in 1991, and then fluctuated at 0.85–0.86 for the next few years. As I will show in the next two chapters, the decline of CVs is related to the fact that the leading provinces of Shanghai, Beijing, Tianjin, and Liaoning experienced slower growth as their economies were tightly controlled by the central state and they had difficulty in adjusting to the emerging market mechanism. On the other hand, Jiangsu, Guangdong, Shandong, Zhejiang, and Fujian have increased their income rapidly through the growth of rural enterprises, domestic and foreign investment, as well as international trade. However, with the recovery of Shanghai during the early 1990s, the trend of declining interprovincial inequality has slowed down.

Such a fluctuating trend of interprovincial inequality can also be found in the change of per capita national income of China's richest and poorest provinces. In 1952, per capita national income of Shanghai was six times higher than the provincial average, whereas the poorest province of Guizhou was only 56 percent of that average. The gap between these two provinces reached its peak in 1978 when Shanghai's per capita national income was seven times higher than the average and that of Guizhou was only 37 percent of that average. During the reform period, however, the leading role of Shanghai has declined drastically.

Without the consideration of the three municipalities, the trend of interprovincial* inequality did not change much from the late 1970s to 1991 (Figs 2.2 and 2.4). However, gaps among these provinces increased during the 1991–95 period. This further confirms that declining interprovincial inequality during the reform period is largely related to the declining status of the three municipalities.

In summary, from 1952 to 1996, the relative positions of the eastern, central and western regions did not change much, with the eastern region's per capita income always above the national average and higher than the other two regions. The relative positions of the eastern region were improved during both Mao's era and the post-Mao period. The late 1950s, late 1960s, 1970s, and early 1990s were the four major periods during which the coastal region developed faster than the inland regions. During the reform period, interregional inequality increased, while interprovincial inequality actually declined. A rapid rise of interregional inequality occurred during 1992–95. In the mid-1990s, the coastal–interior inequality in China became more intense than ever before.

2.5 New forms of uneven regional development

With the change of mechanisms, the relative economic status of regions has also changed over time. During Mao's era and especially influenced by the industrialization policy, traditional economic centers of China, particularly the three municipalities and northeastern provinces, experienced more rapid growth. Since the implementation of reforms, however, these provinces have had difficulty in adjusting to market transition, while state policy and foreign investment have favored the development of other coastal provinces. As a result, the status of traditional industrial bases has declined, while several provinces in coastal China have grown more rapidly. Therefore, a new form of uneven regional development has emerged in post-Mao China.

2.5.1 Regional disparities in growth rates

An analysis of growth rates of per capita national income can shed more light on regional development in China. Growth rates varied considerably during Mao's era as well as during the reform period. In the 1952–78 period, the three municipalities (Beijing, Shanghai, and Tianjin) and Liaoning had the highest per capita national income and achieved the highest growth rates—ranked, respectively, first, second, third, and fourth in the nation (Table 2.3 and Fig. 2.5). On the other hand, the provinces of Anhui, Xinjiang, Neimenggu, and Guizhou (the poorest) experienced the slowest growth, with annual rates of less than 2.5 percent. The rapid growth of richer provinces and the slower growth of poorer provinces indicated that interprovincial inequality rose under Mao.

Table 2.3 Regional per capita national income, 1952–92

	1952	*1978*	*1992*	*1952–78 annual growth (percent)*	*1978–92 annual growth (percent)*
Average	99	315	1712	3.9	7.5
Eastern region	118	404	2333	4.6	8.0
Beijing	265	1044	4620	7.1	6.5
Tianjin	267	1040	3765	5.3	5.4
Shanghai	595	2248	6562	5.9	6.3
Hebei	111	315	1520	3.6	6.5
Liaoning	196	610	2650	5.0	5.7
Jiangsu	96	359	2467	3.8	10.1
Zhejiang	103	290	2539	3.6	11.8
Fujian	96	236	1905	2.7	10.1
Shandong	86	274	1962	4.1	9.2
Guangdong	89	320	2767	3.1	10.8
Guangxi	62	190	1112	2.8	6.5
Central region	96	272	2088	2.9	6.5
Shanxi	95	282	1372	2.9	6.4
Neimenggu	153	257	1396	1.9	6.7
Jilin	145	333	1724	2.7	6.8
Heilongjiang	211	488	1956	2.1	4.8
Anhui	78	222	1148	1.7	7.2
Jiangxi	104	243	1198	2.0	6.7
Henan	76	205	1154	3.4	7.4
Hubei	84	297	1485	3.1	7.2
Hunan	78	248	1230	3.3	5.9
Western region	69	222	1165	3.4	6.9
Sichuan	59	212	1154	3.3	7.2
Guizhou	56	156	810	2.2	6.5
Yunnan	63	203	1173	3.5	7.4
Xizang	—	—	1241	—	—
Shaanxi	77	257	1179	4.5	6.5
Gansu	95	294	1156	3.5	6.0
Qinghai	96	336	1364	4.4	4.0
Ningxia	108	282	1249	2.4	5.3
Xinjiang	159	279	1863	1.9	8.2

Sources: SSB 1990, 1994; see SSB 1991–97.

Note: Per capita data in current prices; growth rates in comparable prices.

During the reform period, however, patterns of regional growth have changed greatly. From 1978 to 1992, the growth rates of per capita national income in Shanghai, Tianjin, Beijing, and Liaoning were among the lowest in the nation, and their relative positions in the national economy declined (Table 2.3). Tables 2.4 and 2.5 show that the three municipalities experienced the largest decline in their economic status, as indicated by the change of location quotients, although the status of Shanghai has been stabilized

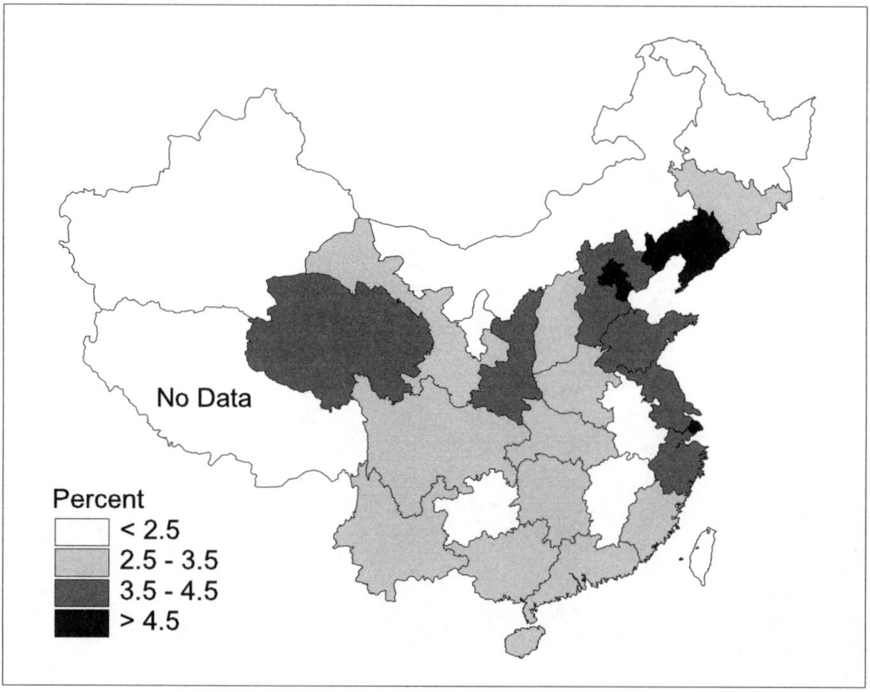

Figure 2.5 Average annual growth rates of per capita national income, 1952–78.
Source: Table 2.3.

during the 1990s. On the other hand, during the 1978–95 period, the coastal provinces of Zhejiang, Guangdong, Jiangsu, Shandong, and Fujian experienced the fastest growth (Table 2.5 and Fig. 2.6). Thus, income gaps between the four traditionally leading provinces (the three municipalities and Liaoning) and the five emerging coastal provinces were reduced. Meanwhile, the poorest provinces of Guizhou, Guangxi, Yunnan, and Sichuan had growth rates higher than the four traditionally richest provinces. As a result, income gaps between the richest provinces and poorest provinces narrowed, leading to the decline of interprovincial inequality, although interregional inequality continued to rise.

During the 1990s, per capita GDP in the five coastal provinces grew by more than 15 percent annually, which ranked them the highest in the nation (Table 2.5). The growth rates of the three municipalities were higher than the national average and the poorest provinces. As a result, the trend of declining interprovincial inequality lessened. Meanwhile, the rapid growth of the five coastal provinces (Guangdong, Fujian, Zhejiang, Jiangsu, and Shandong) stimulated the growth of the coastal region as a whole, leading to the intensification of interregional inequality in China. By 1995, the coastal provinces of China clearly dominated the nation's economy (Fig. 2.7), with

Table 2.4 Location quotients based on per capita national income, 1952–92

	1952	*1960*	*1978*	*1992*	*1952–60 change*	*1960–78 change*	*1978–92 change*
Average	100	100	100	100	0	0	0
Eastern region	119	132	143	153	13	11	10
Beijing	268	358	597	529	90	239	−68
Tianjin	269	349	380	288	80	31	−92
Hebei	112	88	103	91	−24	15	−12
Liaoning	198	338	259	207	140	−79	−52
Shanghai	601	784	994	852	182	211	−142
Jiangsu	97	57	94	132	−40	37	38
Zhejiang	104	86	96	167	−18	10	71
Fujian	97	81	73	102	−16	−8	29
Shandong	87	56	91	114	−31	35	23
Guangdong	90	70	74	112	−20	4	38
Guangxi	63	47	47	41	−16	0	−6
Central region	97	85	76	67	−12	−9	−9
Shanxi	96	92	75	65	−4	−17	−10
Neimenggu	154	142	93	85	−12	−49	−8
Jilin	146	137	108	99	−9	−29	−9
Heilongjiang	214	175	138	97	−39	−37	−41
Anhui	78	52	45	44	−26	−7	−1
Jiangxi	105	75	65	59	−30	−10	−6
Henan	77	63	69	68	−14	6	−1
Hubei	85	74	69	66	−11	−5	−3
Hunan	79	72	68	56	−7	−4	−12
Western region	69	63	62	58	−6	−1	−4
Sichuan	59	48	51	49	−11	3	−2
Guizhou	56	53	37	33	−3	−16	−4
Yunnan	64	60	58	58	−4	−2	0
Tibet	—	—	—	—	—	—	—
Shaanxi	77	89	90	80	12	1	−10
Gansu	96	64	86	71	−32	22	−15
Qinghai	97	109	110	69	12	1	−41
Ningxia	109	78	75	57	−31	−3	−18
Xinjiang	161	158	97	106	−3	−61	9

the economic status of Guangdong, Zhejiang, Jiangsu, Shandong, and Fujian moving up quickly. Guizhou remained the poorest province in the nation. The provinces with GDP per capita less than 3000 yuan also included Gansu, Tibet, Shaanxi, and Jiangxi.

2.5.2 Grouping of the provinces

Based on the change of LQs of per capita national income, especially during the post-Mao period, and for simplicity, I have classified China's provinces into six groups (Table 2.6).[13] A comparison of provincial LQs in 1952, 1978,

Table 2.5 Regional per capita gross domestic product, 1978–95

	Per capita GDP (yuan)						*Location quotient (average = 100)*	
	1978	*1990*	*1995*	*1978–90 annual growth (percent)*	*1990–95 annual growth (percent)*	*1978–95 annual growth (percent)*	*1978–90 change*	*1990–95 change*
Average	359	1511	4792	7.6	14.5	9.0	0	0
Eastern region	462	1964	6777	7.9	14.5	9.8	3	13
Beijing	1289	4754	11741	7.1	9.2	7.7	−22	−46
Tianjin	1150	3582	9804	5.8	10.0	7.0	−61	−27
Hebei	364	1389	4444	6.5	13.2	8.4	−12	3
Liaoning	663	2463	6847	6.7	9.4	7.5	−19	−21
Shanghai	2498	5700	17774	5.9	11.7	7.5	−128	−19
Jiangsu	430	1978	7319	9.0	16.8	11.3	19	28
Zhejiang	329	1998	8185	10.7	18.4	12.9	36	37
Fujian	273	1551	6731	8.7	17.4	11.2	9	20
Shandong	321	1599	5758	8.3	15.1	10.3	7	12
Guangdong	333	2263	7939	10.6	16.9	12.0	35	27
Guangxi	223	922	3535	5.2	15.0	8.0	−14	5
Central region	310	1264	3691	7.1	10.3	8.0	−4	−8
Shanxi	365	1410	3569	6.1	8.4	6.8	−16	−14
Neimenggu	310	1338	3666	7.2	8.4	7.6	−4	−14
Jilin	382	1587	4356	7.7	9.7	8.3	2	−13
Heilongjiang	559	1887	5465	5.0	6.8	5.5	−41	−26
Anhui	242	1089	3348	8.4	12.5	9.6	6	0
Jiangxi	276	1118	2984	7.4	11.9	8.8	−2	−2
Henan	232	1061	3313	8.3	11.4	9.2	5	−3
Hubei	332	1484	4162	7.6	11.2	8.7	−1	−5
Hunan	286	1157	3445	6.2	10.0	7.3	−12	−7
Western region	254	1078	2945	7.5	8.9	7.9	−1	−10
Sichuan	238	1065	3136	7.8	10.0	8.5	1	−7
Guizhou	175	790	1809	7.4	7.0	7.3	−1	−10
Yunnan	226	1073	3044	8.1	8.6	8.2	3	−10
Tibet	375	1101	1908	5.9	7.8	6.4	−19	−16
Shaanxi	294	1148	2859	7.0	7.9	7.2	−6	−14
Gansu	348	1058	2298	6.6	7.8	7.0	−11	−16
Qinghai	426	1481	3437	4.7	6.0	6.0	−34	−21
Ningxia	353	1321	3338	6.9	6.1	6.1	−8	−22
Xingjiang	316	1689	5069	8.9	9.6	9.1	12	−12

Source: SSB 1990–96, 1996c.
Note: Per capita data in current prices; growth rates and location quotient in comparable prices.

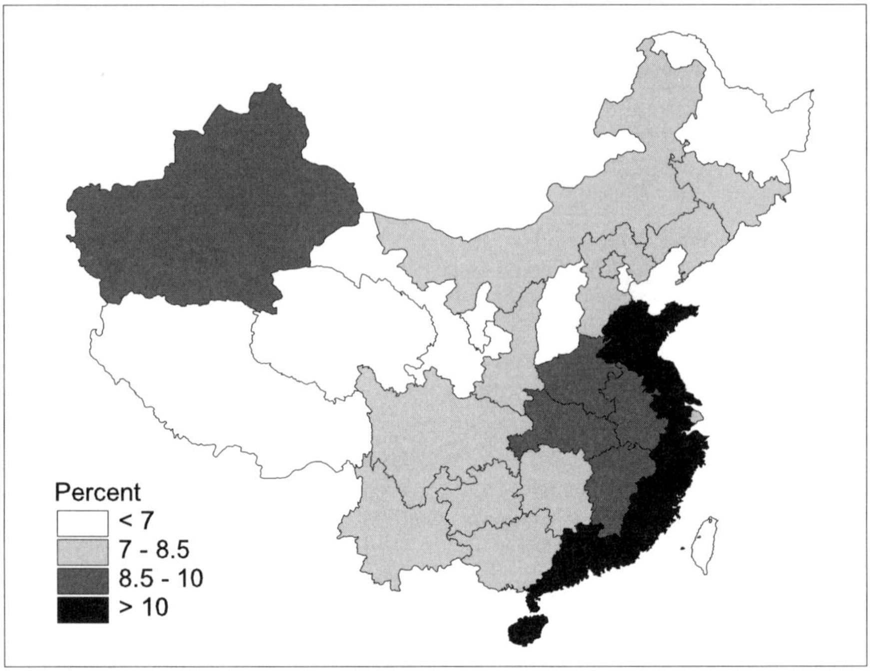

Figure 2.6 Average annual growth rates of per capita GDP, 1978–95.
Source: Table 2.5.

and 1992 clearly shows the changing forms of regional development. From 1952 to 1978, the LQs of the provinces in Group I (Shanghai, Beijing, and Tianjin) and Liaoning (Group II) increased rapidly, indicating that during Mao's era their economies became wealthier than other provinces. The LQs of other coastal provinces, except Shandong, declined, particularly during 1952–60. The provinces losing the most in LQs also include the interior provinces of Xinjiang, Neimenggu, Jiangxi, Ningxia, and Anhui.

From 1978 to 1992, the LQs of the three municipalities and northeastern industrial provinces (Groups I and II) declined the most while the provinces in Group III increased rapidly. LQs of most provinces in Groups IV–VI declined. During the 1990s, the upward trend of the coastal provinces accelerated, while the positions of most traditional industrial bases and poor interior provinces lost ground further. The following section will demonstrate these changing forms of uneven development, which will be explained more thoroughly in Chapter 4.

2.5.3 Development paths of the provinces

The change of LQs shows that during the last four decades the status of the traditionally leading economic bases (Group I) rose under Mao but

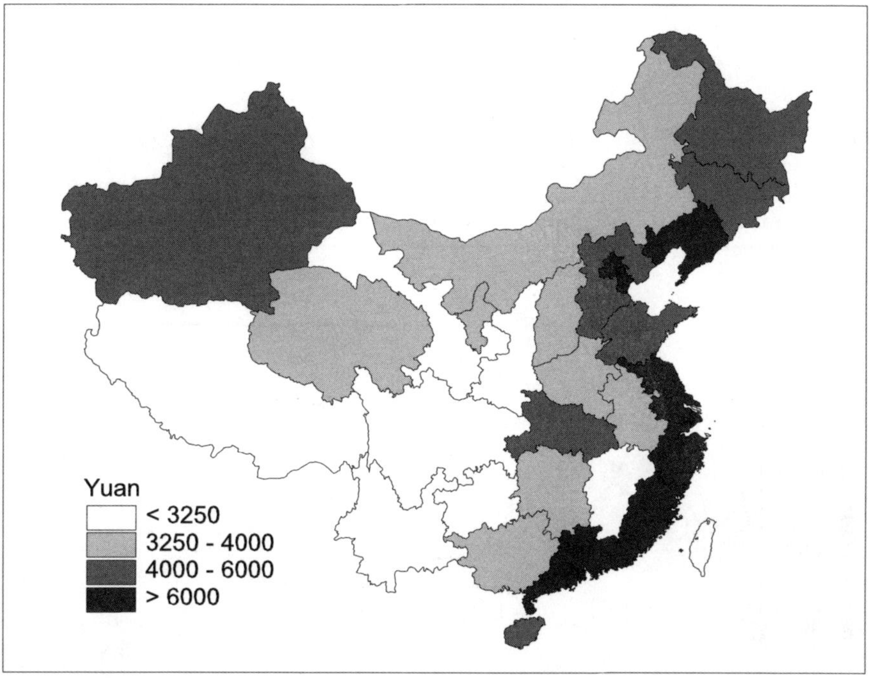

Figure 2.7 Per capita GDP, 1995.
Source: SSB 1996.

declined during the reform period, although their income levels have always been among the highest in China (Fig. 2.8). It is interesting to note that these municipalities became more affluent during the 1950s. During that decade, as these provinces were China's major industrial and political centers, the central government provided them with preferential policies. Although their status declined in the early 1960s when Chinese cities were severely damaged due to the GLF, these provinces recovered quickly and their relative positions in the Chinese economy were improved.

Since the launch of economic reforms, however, the relative positions of these provinces declined steadily. From 1978 to 1990, the growth rates of Shanghai, Beijing, and Tianjin were among the lowest in the nation. The slowdown in growth was related to the drastic reduction of state-subsidized energy and raw materials, the tight central control over local budget, the dominance of SOEs, and the slow start of the 'open door' process. The economic competitive edge that these provinces enjoyed for years through large-scale state support of the urban economy was eroded swiftly by the rapid growth of several coastal provinces. The declining status of these municipalities in the national economy eventually caught the attention of

Table 2.6 Grouping of provinces based on location quotients

Group		*Provinces*
I	Urban	Shanghai, Beijing, Tianjin
II	Industrial	Liaoning, Jilin, Heilongjiang, Shanxi, Hebei
III	Coastal	Jiangsu, Shandong, Zhejiang, Fujian, Guangdong
IV	Central	Hunan, Hubei, Henan, Anhui, Jiangxi
V	Southwestern	Guizhou, Yunnan, Sichuan, Guangxi, Tibet
VI	Northwestern	Shaanxi, Ningxia, Gansu, Qinghai, Xinjiang, Neimenggu

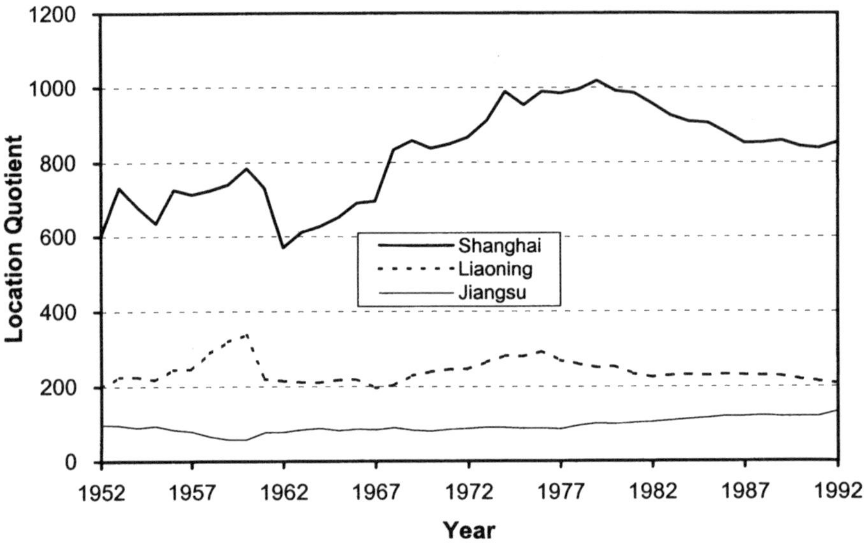

Figure 2.8 Trends of location quotients based on per capita national income: Groups I, II and III.

the central government. Since the early 1990s, their growth rates have been accelerated and their status in the national economy is recovering.

Before 1949, the provinces of Group II were among those with the highest per capita industrial output. During Mao's era, their leading roles in China's development were not challenged significantly, due largely to Mao's emphasis on industrialization through the development of heavy industry. During the First FYP, these provinces were highlighted in the development of heavy industry because of their rich resources and structural foundation of industry. A large number of investment projects, particularly in the iron and steel industry, were allocated to those provinces. However, they have been losing their competitive edge in the post-Mao period, and their growth rates were among the lowest in the nation. From 1978 to 1992, the annual growth rate of per capita national income for Liaoning was 5.7 percent, the second lowest among the coastal provinces (only higher than Tianjin). During the 1990–95 period, the annual growth rate of per capita GDP was 9.4 percent in Liaoning, only slightly higher than Beijing (9.2 percent). The north-eastern provinces of China experienced slower growth and have gradually lost the strong economic positions that they enjoyed in Mao's era. But some preferential policies have been provided to restructure their economies, such as increasing investment in renovating old industrial equipment. Although the restructuring process is slow, Liaoning is gradually adjusting to the emerging market mechanisms.

The coastal provinces (Group III) were not favored by Mao's regime, but they are the biggest winners of economic reforms. During the 1950s, with the implementation of the First FYP that favored the interior and the traditional industrial bases, the status of these provinces in the national economy declined greatly. Devoid of the preferences of the central government, these provinces developed strong abilities to use local resources and markets to improve their economies. Their status in the national economy began to recover after the early 1970s when a pragmatic leadership emerged and a decentralization policy was implemented in China. During the reform period, these provinces have registered the fastest income growth. Their status has been advanced dramatically, with Zhejiang, Guangdong, and Jiangsu moving up most rapidly, followed by Fujian and Shandong. They have actively bargained with the central government for more decentralized policy and for allowing them to initiate bolder local policies. Moreover, SEZs in Guangdong and Fujian were opened to foreign investment and trade in 1980, which was soon followed by the opening up of cities in other coastal provinces. Meanwhile, regions have aggressively launched new initiatives to stimulate local economic growth. Rural enterprises and the service sector have flourished there and have become the fastest growing sectors. All these changes have contributed to the region's phenomenal growth in the reform period, especially in the 1990s with the implementation of bolder reform policies.

Among the provinces of Group IV, the positions of Hunan, Hubei, and Henan were relatively stable during much of Mao's era, indicating that they had somewhat benefited from interior development policy (Fig. 2.9). However, the positions of peri-coastal provinces (Anhui and Jiangxi), among the poorest in the nation, declined, particularly after the late 1960s. These two provinces were largely ignored in investment allocation during Mao's era. During the early period of reforms, the status of the provinces of Group IV declined further, particularly with the emphasis of economic reforms shifting to urban and industrial reforms in the mid-1980s. However, during the 1990s, with the efforts of these provinces and the extension of economic reforms from the coastal provinces to the interior, their positions in the national economy have been rising gradually.

The income of the traditionally poor provinces (Group V) has remained low. Though their status was improved in the early 1950s and early 1960s

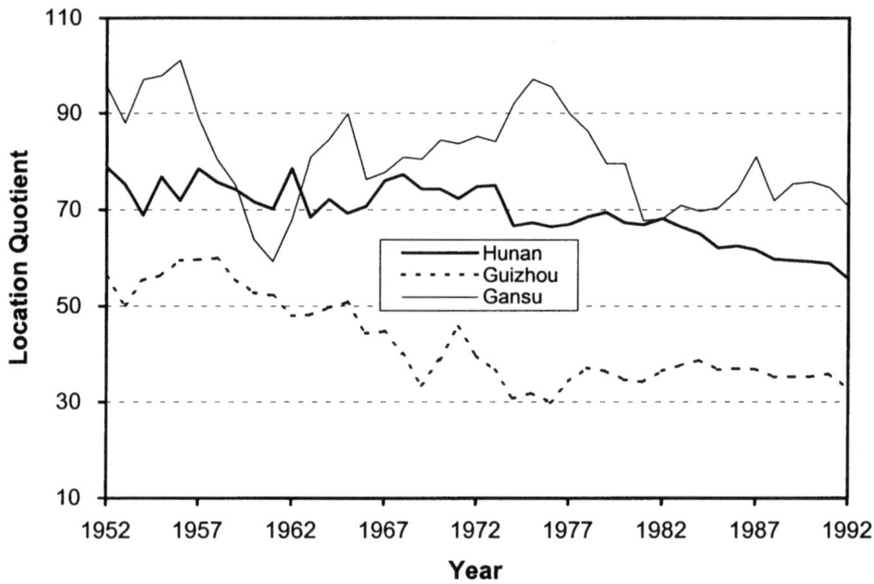

Figure 2.9 Trends of location quotients based on per capita national income: Groups IV, V and VI.

when Mao attempted to develop China's backward interior regions, the overall LQs under Mao declined, particularly after the early 1960s. As shown in Fig. 2.9, Guizhou, the poorest province in China, also experienced a trend of declining status during Mao's era. It is clear that the relative conditions of the poorest provinces did not improve at all from the development policy under Mao. In the post-Mao period, the status of the provinces in this group has dropped further. They had difficulty in competing with the coastal provinces and in adjusting themselves to the emerging market economies. The biggest slip in the LQs took place in the early and mid-1990s when more rapid market transition was taking place in China, and their income gaps with the coastal provinces intensified.

The status of the provinces of Group VI fluctuated dramatically during Mao's era. Some improvement was made in Xinjiang and Neimenggu during the 1950s. However, with the breakup of Sino–Soviet relations in the early 1960s, these two provinces were at the forefront of Sino–Soviet border conflicts. A perceived invasion from the Soviet Union in the north led to the decrease of state investment in these provinces. Their economic status declined rapidly during the 1960s and 1970s. During the reform period, the status of this group has declined further, although Xinjiang was relatively better off during the 1980s. The most rapid decline was recorded during the 1990s, a pattern similar to many other interior provinces.

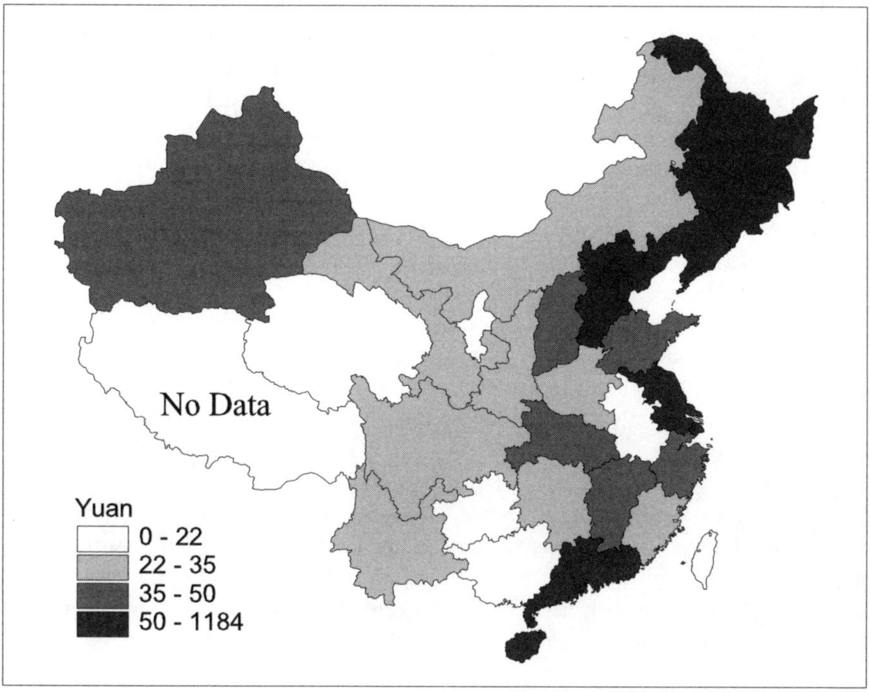

Figure 2.10 Per capita gross value of industrial output, 1952.
Source: SSB 1990.

2.5.4 Regional patterns of industrial production

Industrialization and regional industrial allocation were the focus of socialist development policy. During Mao's era, the Chinese government emphasized industrialization to achieve rapid economic growth and equal distribution. Mao (1956) directed that more investment be allocated to the interior so that 'industry may become evenly distributed and rationally sited over the whole country' (p. 66). In addition, industry has always received the lion's share of state investment and yielded the bulk of national production. In 1990, the industrial sector absorbed 55.9 percent of capital construction investment in SOEs and produced 45.8 percent of national income. Analyzing industrial distribution among regions should help to further understand changing patterns of regional development and inequality in China.

In 1952, the eastern region, with 14 percent of the total area and 42.3 percent of the population, produced 68.1 percent of gross value of industrial output (GVIO). The three municipalities, the three northeast industrial provinces and Jiangsu registered the highest per capita GVIO (Fig. 2.10). Industries were poorly developed in the interior provinces, especially in Qinghai, Ningxia, Anhui, and the southwestern provinces of Yunnan, Guizhou, and Guangxi.

The First FYP (1953–57) began to emphasize interior development for the purpose of rapid industrialization, regional equality and national defense. As a result, interregional and interprovincial inequalities decreased during the 1950s. This suggests that the nation's development policies in the First FYP narrowed the gaps of industrial production among China's regions and provinces. For the next two decades, however, regional industrial inequality persisted, and the eastern region outgrew the other two regions. During this period, the defense-oriented 'Third Front' program was implemented and a large percentage of state investment was allocated to the interior, but these projects were costly and yielded poor economic returns. Meanwhile, interior policies favored only a few provinces. China's remote and minority provinces were largely neglected and did not receive much benefit from interior development policies. In addition, during the early 1970s, China gradually opened its door to foreign countries and normalized its relations with the United States. Most of the large-scale capital-intensive industrial equipment imported from the Western countries were allocated to major cities located on the coast or near the Yangtze River, and TVEs were established in the coastal provinces. In 1978, the northeastern provinces, the three municipalities and Jiangsu still had the highest per capita GVIO, and patterns of industrial distribution did not change much from that of 1952, although a few provinces in northwestern China improved their status (Fig. 2.11).

During the reform period, patterns of regional industrial inequality behaved broadly similar to that based on national income. From 1978 to 1990, the eastern region experienced faster growth than the central and western regions. Growth rates of Zhejiang, Guangdong, Jiangsu, Fujian, and Shandong were much higher than other provinces, while that of the three municipalities and Liaoning were among the lowest. Shanghai's industrial growth rate ranked the lowest in the nation (6.5 percent). From 1952 to 1990, the share of industrial output produced by the three centrally administered municipalities (Shanghai, Beijing, and Tianjin) and the northeastern industrial bases (Liaoning, Jilin, and Heilongjiang) decreased from 48.8 percent to 25.3 percent, especially Shanghai where it decreased from 19.4 percent to 6.8 percent. The persistence of interprovincial inequality during the 1960s and the decline after the economic reforms correspond well with the changing status of these provinces. During the 1990s, the coastal region still outgrew the central and western regions. The five coastal provinces again recorded the highest growth rates, while Shanghai's growth rate recovered somewhat. Growth rates of China's traditional industrial bases, including Jilin, Heilongjiang, and Tianjin, were still among the lowest. Several of the poorest provinces in the interior region also experienced slower growth. By 1995, per capita GVIO in the eastern region was three times as high as central and western regions. Except for Guangxi, eastern provinces had higher per capita GVIO than any central and western provinces (Fig. 2.12). The eastern region has clearly dominated the nation's industrial production.

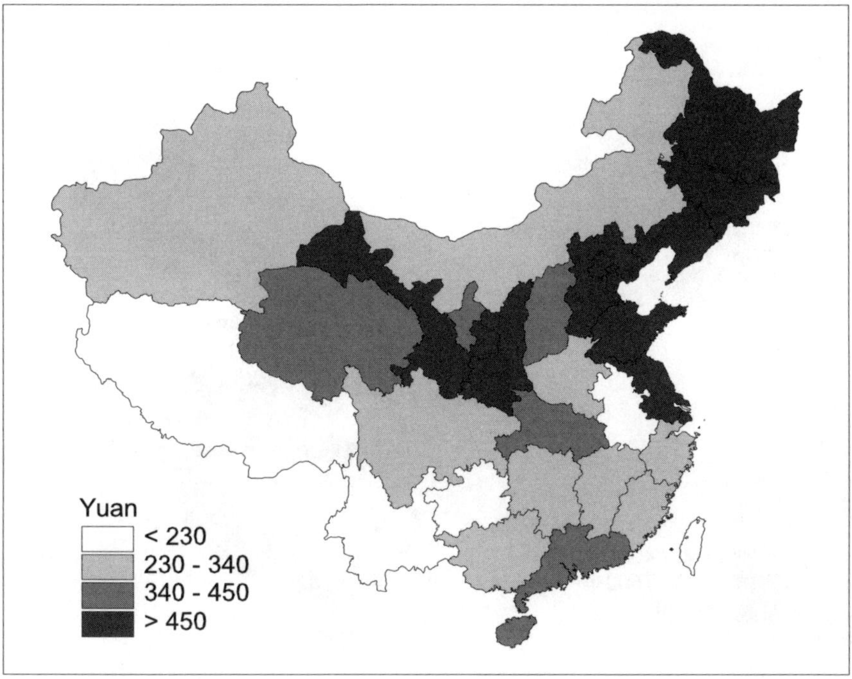

Figure 2.11 Per capita gross value of industrial output, 1978.
Source: SSB 1990.

2.6 Conclusion

Regional inequality has been an important issue to Chinese leaders since the founding of the PRC. Some efforts were made by the government to allocate resources in a spatially more balanced way to reduce regional inequality, especially during the First Five-Year Plan period. However, from the late 1950s to the mid-1970s, little consideration was given to the development of the poorest provinces. During the 1952–78 period, both interregional and interprovincial inequalities based on per capita national income increased. Overall, China's regional inequality remained substantial during Mao's era. Traditional economic centers recorded more rapid growth, while the status of poorer interior provinces did not improve.

Since economic reforms were launched in 1978, rapid growth and dramatic changes have taken place in China. Uneven regional development based on comparative advantage was considered as necessary to achieve rapid and efficient economic growth. A spatially balanced pattern of development is no longer emphasized, although the idea has never been totally discarded and has been re-emphasized lately. Interregional inequality

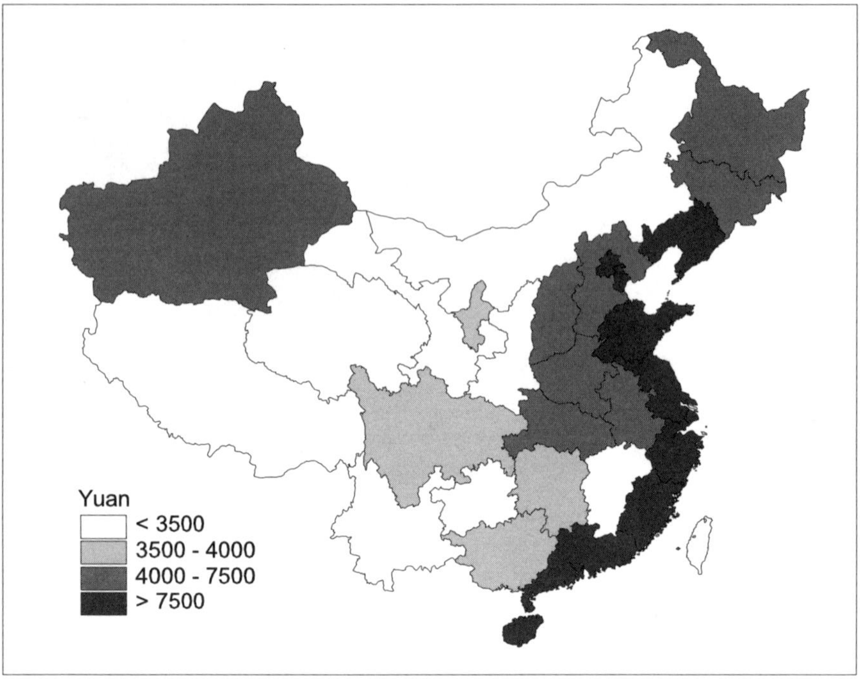

Figure 2.12 Per capita gross value of industrial output, 1995.
Source: SSB 1996.

has risen further, and its increase during the 1991–95 period was substantial. The core–periphery structure established during China's semi-colonial period has persisted after four and a half decades of socialism. Interprovincial inequality, however, declined. This is related largely to the slow growth of the four traditionally leading provinces (Shanghai, Beijing, Tianjin, and Liaoning) and the rapid growth of the five emerging coastal provinces (Guangdong, Jiangsu, Zhejiang, Shandong, and Fujian). In the following two chapters, I attempt to reveal the factors and processes underlying such changing patterns of regional development in China.

3 Fiscal systems, investment, and regions

In this and the next chapter, I will focus on the analysis of the multiple mechanisms and agents underlying China's regional development, especially the central state, local states and factors, and foreign investment. Centralized fiscal and investment systems, in which central governments controlled resource allocation, were critical to socialist development policies. The failure of the orthodox socialist system prompted socialist countries to reform their fiscal and investment systems. Fiscal and investment reforms are among the major components of reforms, and exert a great impact on regional development in China.

Investment is a key factor of production under both capitalist and socialist systems. It is even more important to socialist countries, as investment policy is the central element of socialist development strategies and socialist systems tend to operate under 'investment hunger' (Kornai 1980, 1992; Naughton 1995a). However, the characteristics, determinants, and impact of investment in these countries remain poorly studied, although a number of publications have examined regional investment in developed countries (e.g., Gertler 1986; Florida and Smith 1993). Investment plays an important role in the Chinese economy, and analyzing investment policy and regional investment should help understanding of China's regional development.

This chapter investigates the temporal changes of fiscal and investment systems in China, analyzes regional patterns of fiscal and investment structures and discusses the impact of fiscal and investment changes on regional development. Unlike studies on central–local fiscal relations that emphasize revenue sharing, my analysis investigates both revenues and expenditures and pays special attention to regional disparities. Moreover, I emphasize the analysis of regional investment as a key factor of economic and regional growth in China. I also highlight the changing patterns and underlying factors of investment in order to properly understand patterns of regional development.

3.1 Fiscal systems and regions

3.1.1 Fiscal systems and regions under Mao

During Mao's era, China, like the former Soviet Union, had a centralized and hierarchical economic system. This system was complicated, but mainly consisted of centralized fiscal systems and budgeting process, a monopolized People's Bank of China (PBC), and a complicated investment and material allocation system. However, Chinese economic and planning systems changed frequently, corresponding to the change of policy environments and strategic goals. Economic planning was often poorly made and poorly implemented, especially during the Cultural Revolution.

To control inflation and influenced by the Soviet Union, China adopted a system of total fiscal centralization in 1950, shortly after the establishment of the new republic in 1949 (Table 3.1). This system was featured by the turning over of all local revenues to the center, the distribution of all local expenditures by the center, and the lack of correspondence between revenues and

Table 3.1 China's fiscal systems under Mao, 1950–76

1950	Total centralization (*tongshou tongzhi*) 1. All local revenues turned over to the center 2. All local expenditures disbursed by the center 3. Lack of correspondence between revenue and expenditure
1951–57	'Multilevel management, revenue-sharing by category and fixed rate' (*tongyi lingdao, fenji guanli, shouru fenlei fencheng*) 1. Three levels of budgetary administration 2. Revenue sharing
1958	Decentralized program 1. New revenue sources for localities 2. Normal expenditure and extraordinary expenditure 3. Fixed five-year sharing rates
1959–67	'Sharing total amount at annually adjusted rate' (*dingshou dingzhi, shouzhi guagou, zonge fencheng, yinian yibian*) 1. Revenue sharing 2. Annual adjustment 3. Interregional transfer
1968–76	'Cultural Revolution' and fiscal turbulence 1. 1968: Total centralization (the 1950 model) 2. 1969–70: Modification of the 1959–67 model 3. 1971–73: 'Fixing revenue and expenditure, contracting surplus and deficits, annual adjustment' 4. 1974–75: 'Fixed revenue retention rate, sharing above plan surplus, contracting expenditure' 5. 1976: Modification of the 1959–67 model

Source: Editorial Committee 1988, Oksenberg and Tong 1991.

expenditures. Between 1951 and 1957, the fiscal system remained centralized in order to finance the Korean War, to implement the Soviet-style industrialization, and to maintain economic and political stability. Fiscal decentralization took place in 1958 when Mao initiated the Great Leap Forward which soon led to economic and financial crises. The provinces obtained fixed five-year sharing rates with the center and received new resources and more responsibilities in expenditures.

China experienced fiscal recentralization during the 1960s to adjust the economy in the early 1960s and to carry out defense projects in the mid- and late 1960s. Fixed sharing rates were changed to annually adjusted rates, which gave the provinces less leeway in retaining increased revenues as the center could claim a larger share of the increased revenues. Revenues collected in excess of approved expenditures had to remit to the center while the provinces with revenues less than expenditures received transfers from the center.

Total centralization was reintroduced in 1968, when the economy of the nation went into crisis with the initiation of the Cultural Revolution. The year 1970 witnessed the initiation of a more decentralized fiscal system, followed by decentralization policies in the early 1970s. Under the 1971–73 system, the provinces remitted a fixed sum of revenues to the central government, and retained savings in expenditures and the revenues after the lump-sum transfer. Moreover, most centrally owned enterprises were shifted to local management, which provided more revenues for the provinces. The system of decoupling expenditure from revenue was implemented in 1974–75 to regain the central control over budget. The center made appropriations for provincial outlays regardless of whether the province met the revenue collection target. The provinces could only retain a small share of the local revenue and arrange local spending under a uniform ceiling. The 1976 system was further centralized to strengthen the control over regions under unstable political and economic conditions.

Several major features of Mao's fiscal system and central–provincial fiscal relations can be summarized. First, the central government enjoyed considerable power in the budget process with little role played by the provinces. Fiscal planning occupied a pivotal role in China's Soviet-style central planning. The Chinese budget was 'consolidated' to include the budgets of all administrative levels of the government (Wong *et al.* 1995). The budget process was tightly controlled by the central government, which directed that each province is as a chess piece in a game of chess and must follow the principle of 'all the nation is a chess-game' and 'unified leadership, level-by-level-management.'[1] Fiscal decentralization took place in the late 1950s and the early 1970s, but both resulted in fiscal chaos, and were followed by recentralization.

Secondly, the execution of provincial budgets was monitored by the PBC and local economic and financial departments controlled by the center. The provinces were required to collect revenues as the agents of the central government (Tong 1991). The center not only set ideological guidance for the provinces to follow but also established a series of administrative mechanisms to control them. For example, it had control over the appointment, promotion, transfer and removal of the top officials in the provincial party. The Ministry of Finance approved the annual revenue and expenditure plans of the provinces. But the conflict between central control and local autonomy never stopped.

Thirdly, the correspondence between revenue and expenditure was weak. The central government transferred revenues collected from budget-surplus provinces to budget-deficit ones and to the provinces with strategic significance, such as the provinces where the defense-oriented projects were constructed. This transfer was also used to equalize regional income and social services. Through the mechanism of transfer, the central government partly redistributed funds from the affluent provinces to the poor provinces and strategic locations; but a tremendous amount of revenue was wasted in this transfer process.

The centralized fiscal system partially enabled the central government to exert control over the provinces and to mobilize capital among sectors and regions. It played a pivotal role in the construction of the 156 key projects imported from the former Soviet Union during the First Five-Year Plan (FYP) period, the construction of the Third Front project from the mid-1960s to the early 1970s, and in the center's effort to retrench capital investment and to adjust the national economy during the early 1960s.

However, Mao's fiscal system was troubled by many serious problems. First, the overly centralized budget system made the regional economies sensitive and extremely vulnerable to national events. Tensions arose from revenue sharing and inadequate planning as local factors were given little consideration in the planning process. Secondly, the provinces lacked incentives to stimulate economic growth as a result of the weak correspondence between revenue and expenditure. Vertical (*tiao*) and horizontal (*kuai*) government organizations lacked coordination in fiscal policies.[2] Thirdly, much of the revenue transferred interregionally was wasted in inefficient projects while budget-surplus provinces lacked the necessary funds for economic development. Lastly, fiscal decentralization often led to economic mismanagement by regions, as a well-coordinated adjustment system was lacking.[3]

3.1.2 Post-Mao fiscal reforms and regions

Problems with Mao's systems prompted post-Mao economic reforms in order to improve economic efficiency and to revitalize China's stagnant economy. Fiscal reforms attempted to give the provinces more fiscal autonomy and to stimulate revenue generation, but fiscal policies varied greatly among the provinces. The 1977–79 period was marked by a series of fiscal reform experiments (Table 3.2). The Jiangsu experiment, initiated in 1977, introduced

Table 3.2 Post-Mao China's fiscal systems, 1977–95

1977–79	Fiscal experiments 1. The Jiangsu and Sichuan models 2. The Guangdong and Fujian experiments 3. Experiments with different revenue-sharing schemes
1980	'Dividing revenues and expenditures, multilevel contracting' (*huafen shouzhi, fenji baogan*) 1. 'Eating in separate kitchens' (*fenzao chifan*) 2. Fixed five-year revenue-sharing rates 3. Fiscal decentralization: Five central–provincial fiscal systems
1985	'Dividing tax categories, checking revenues and expenditures, multilevel contracting' (*huafen shuizhong, heding shouzhi, fenji baogan*) 1. Three types of revenues: central fixed, local fixed, and shared 2. Specific revenue sharing 3. Three central–provincial fiscal systems
1988	Five central–provincial fiscal systems 1. Basic sharing 2. Basic sharing with growth adjustment 3. Fixed quote delivery (fixed-sum transfer) 4. Fixed quote with growth adjustment 5. Fixed central subsidies
1994	Tax assignment system (*fen shui zhi*) 1. Central taxes 2. Local taxes 3. Shared taxes

Source: Oksenberg and Tong 1991; Agarwala 1992; Xiang *et al.* 1992.

four-year fixed revenue sharing with a local retention rate of 42 percent. Jiangsu was permitted to make decisions on its expenditures if the budget was balanced. The Sichuan model divided revenues into central, provincial, and shared categories and classified expenditures into central and provincial ones. In addition, the central government reached contract-sharing agreements with six other provinces (Hubei, Hunan, Zhejiang, Fujian, Hebei, and Beijing).

In 1978, ten provinces carried out an experimental system in which the provinces and the center shared increased revenue, and provincial revenue was tied to provincial expenditure. These provinces obtained fixed three-year local revenue retention rates with the center: Shaanxi (75 percent), Zhejiang (55 percent), Hunan (55 percent), Gansu (70 percent), Jiangxi (70 percent), Fujian (75 percent), Shandong (40 percent), Jilin (70 percent), Heilongjiang (45 percent), and Beijing (40 percent) (Oksenberg and Tong 1991). In 1979, the above fiscal system was changed to 'sharing of above-planned revenue' to readjust and consolidate the overheated economy. Fiscal experiments were also carried out in Guangdong and Fujian in 1979, which gave them greater fiscal autonomy.

Table 3.3 Variations of provincial revenue-retention rates, 1980–91

	1980	*1982*	*1985*	*1988*	*1991*
Eastern region					
Beijing	28.1	35.5	48.2	50.0	50.0
Tianjin	30.6	34.9	39.5	46.5	46.5
Hebei	27.2	67.5	69.0	70.0	70.0
Liaoning	30.1	30.0	51.1	58.3	58.3
Shanghai	8.6	10.5	26.0	(10500)	(10500)
Jiangsu	39.0	38.0	39.0	41.0	41.0
Zhejiang	13.0	56.0	55.0	61.5	61.5
Fujian	(150)	(150)	[235]	[50]	[50]
Shandong	10.0	48.9	59.0	(290)	(290)
Guangdong	(1000)	(1000)	(772)	(1410)	(1410)
Guangxi	[297]	[359]	[716]	[610]	[860]
Central region					
Shanxi	57.9	75.4	97.5	87.6	87.6
Neimenggu	[1168]	[1414]	[1783]	[1840]	[2160]
Jilin	[300]	[300]	[397]	[110]	[110]
Heilongjiang	[887]	[887]	96.0	(300)	(300)
Anhui	58.1	77.0	80.1	77.5	77.5
Jiangxi	[138]	[138]	[239]	[50]	[50]
Henan	75.4	82.0	81.0	80.0	80.0
Hubei	44.7	69.0	66.5	100.0	100.0
Hunan	42.0	75.0	88.0	(800)	80.0
Western region					
Sichuan	72.0	85.0	89.0	100.0	100.0
Guizhou	[526]	[636]	[743]	[740]	[900]
Yunnan	[330]	[525]	[637]	[670]	[710]
Xizang	[496]	[601]	[750]	[900]	[910]
Shaanxi	88.0	100.0	[270]	[120]	[120]
Gansu	53.2	80.0	[246]	[130]	[130]
Qinghai	[402]	[486]	[611]	[660]	[740]
Ningxia	[300]	[363]	[494]	[530]	[590]
Xinjiang	[895]	[1083]	[1450]	[1530]	[1760]

Source: Oksenberg and Tong 1991; Agarwala 1992; Bahl and Wallich 1992.

Notes

[ ] Retains all revenue and receives fixed amount from the center (in million yuan).

( ) Pays fixed amount to the center (in million yuan).

These experiments of revenue sharing and fiscal decentralization provided more incentives for the provinces and resulted in significant revenue growth for both the center and the provinces. Based on these experiments and pressure from provincial officials for greater autonomy, the central government decided in 1979 to launch national fiscal reforms in 1980. The 1980 fiscal reform provided the provinces with more fiscal powers in revenue sharing and expenditure. Its fixed five-year revenue sharing stimulated the efforts of the provinces to generate more revenue. The 1980 fiscal system shows dramatic regional variations (Table 3.3):

1 The greatest fiscal autonomy was enjoyed by Guangdong, which remitted a lump-sum to the center for five years, and by Fujian, which received a grant from the center for five years.
2 Jiangsu continued four-year fixed revenue sharing with some adjustment.
3 Fifteen provinces shared specific revenues with a five-year fixed rate with the central government. Sources of revenues were classified into three categories: central fixed, local fixed, and fixed-rate shared revenues. The benefit of increased revenue was divided between the center and the provinces.
4 Eight minority and border provinces (Guangxi, Guizhou, Yunnan, Tibet, Ningxia, Neimenggu, Qinghai, and Xinjiang) not only retained most locally collected revenues and all revenue increments but also received transfers from the center. However, these provinces were not as autonomous in fiscal management as many other provinces.
5 The three centrally administered municipalities of Beijing, Tianjin, and Shanghai shared total revenues with the center and were tightly controlled by the central government. Their revenue-sharing rates were adjusted every year; thus revenue increase that resulted from local fiscal efforts may lead to lower retained revenue-sharing rates in the following years. This arrangement provided the least incentives for revenue collection.

These fiscal reforms were largely implemented but some adjustments were made during 1982–83, due largely to the decline of central revenue, the problem of deficit, and rapid economic changes.

With the initial success of the rural reform and opening up to the outside world, the central government adopted the 'Decision on Economic Structural Reform' in October 1984 and launched a series of reform programs in 1985. A new fiscal system was introduced in 1985, which was designed to address the problem of budget deficit and to facilitate urban reforms. There were three central–provincial fiscal systems:

1 contracting system for Guangdong and Fujian;
2 special fiscal treatments for provinces with large minority groups where they retained all revenues collected and received a fixed percentage of subsidy from the center; and
3 specific revenue sharing for all other provinces.

Sources of revenues were divided into three categories: central fixed, local fixed, and shared. However, this new fiscal system was not well implemented, due largely to the vague divisions of the central, provincial and shared revenues. Many provinces continued the previous systems with some changes.

Some provinces with low revenue-retention rates, such as Shanghai, experienced declining revenues.

The Chinese central government considered the nation to be in the 'primary stage of socialism' and decided to deepen economic reforms in October 1987. This was followed by the 1988 fiscal reform that reduced provincial transfers to the center in order to provide more incentives for local revenue collection. In particular, the three municipalities 'bargained down' the center in revenue sharing and significantly reduced their upward revenue transfers (Table 3.3). Five central–provincial fiscal systems were introduced (Agarwala 1992):

1. Basic sharing in which the provinces remitted a fixed proportion of revenue to the center (Tianjin, Anhui, and Shanxi).
2. Basic sharing with growth adjustment in which a specified nominal proportion was contracted to the center with a certain growth rate (Beijing, Liaoning, Hebei, Jiangsu, Zhejiang, and Henan).
3. Fixed-sum transfer (fixed quote delivery) to the center with no annual adjustments in Shanghai, Shandong, and Heilongjiang where transfers to the center used to be high.
4. Fixed quota with growth adjustment in which a specific amount was contracted to the center with a certain growth rate, including Guangdong and Hunan where the center's share in revenues was small.
5. Budget-deficit provinces received fixed subsidies from the center (16 provinces) (fixed central subsidies).

Several features of post-Mao China's fiscal systems and central–provincial fiscal relations can be summarized. First, the control of budget over the economy has declined, as evidenced by the decrease of the share of budget revenue to national income from 37.2 percent in 1978 to 21.8 percent in 1991. Secondly, fiscal decentralization has given the provinces considerable budget authority. Both budgetary and extrabudgetary funds increased and the center's control over local expenditures was reduced.[4] Thirdly, revenue has been more closely tied to expenditure, and this has significantly reduced interregional income transfer. Finally, the provinces differed considerably in their revenue-sharing arrangements with the central government.

However, whereas post-Mao fiscal systems displayed some characteristics of planned economies, they also created new problems. First, revenue-sharing rates and expenditure ranges depended on negotiation and bargaining between the center and the province. Intense central–local conflicts arose from bargaining over revenue-sharing rates and expenditures. Secondly, fiscal decentralization increased administrative interference in enterprise management (Shirk 1993). The state relied on political instruments to manage the economy. Decentralization was not accompanied by a well-coordinated, market-oriented macroeconomic control system. Thirdly, budget deficit and inflation rates rose sharply, due largely to 'investment hunger'—a phenomenon common to state socialism as described by Kornai (1980)—and to the increase of money supply to reduce budget deficit (Yenal 1990). Rising inflation became one of the most serious problems, leading to political crisis and social instability in the late 1980s (Baum 1994). Fourthly, to meet expenditure needs and to reduce deficit, both central and local governments increased public debts, issued more bonds, and raised administrative charges. Price subsidies also increased dramatically, including that for agricultural products, urban wages, international trade, and food for the urban residents. In 1991, debts, enterprise subsidies, and price subsidies reached 24.68, 51.02 and 37.38 billion yuan, respectively. The proliferation of numerous fees and levies placed mounting pressures on SOEs that were troubled by the problem of low efficiency. Lastly, fiscal systems differentiated the provinces and created unfair competition among regions, stimulating regional conflicts and local protectionism (Wong 1991). These problems well resemble the fiscal problems facing Russia and Eastern Europe arising from revenue sharing and fiscal decentralization (Bird and Wallich 1994).

Tensions and problems arising from fiscal systems prompted the Chinese leaders to push for further fiscal reforms in 1993. A new fiscal system was fully implemented in 1994 after the experiments of 1992–93 in four provinces (Tianjin, Liaoning, Zhejiang, and Xinjiang) and five cities (Shenyang, Dalian, Qingdao, Wuhan, and Chongqing). This reform attempted to lessen the annual intense bargaining with local governments and to ease local antagonism against some coastal provinces such as Guangdong and deep resentment in three centrally administered municipalities, particularly Shanghai, for heavy central control and low local revenue retention rates. More important, the central government attempted to increase the central control of budget revenue and to improve the role of public finance in macroeconomic management. The new fiscal system separates revenue sources from expenditure responsibilities between the center and the provinces. It is a tax assignment system (*fen shui zhi*) which divides taxes into central, local, and shared taxes collected by central or local fiscal agencies; thus, it specifies fiscal responsibilities of central and provincial governments more clearly and replaces traditional bargaining in determining central–provincial revenue-sharing rates.

This new fiscal system has changed the major features of the traditional fiscal systems with some favorable results: it has been reported that revenues have increased rapidly since its implementation. In particular, the shares of central revenue as percentages of national income and total revenues have increased; for example, the center's share of total revenue increased from 22 percent in 1993 to 50 percent in 1995. However, bargaining between central and provincial governments remains intense, and divisions between central and local fiscal responsibilities still need to be improved. With the increase of central revenue, the poor regions expect to receive more transfers from the central government while some rich coastal provinces have been against fiscal recentralization (Chung 1994–95). In short, the impact of this new fiscal system is also regionally uneven and further reform of fiscal systems is needed.

3.1.3 Financial systems, banking, and regions

China's financial systems have also experienced dramatic restructuring. Mao's financial system was formulated in the early 1950s, and was characterized by highly centralized financial institutions, prohibition of financial instruments, repressed interest rates and inflation, and isolation from the world market. The government was the main saver, and the banking system only consisted of the PBC and some rural credit cooperatives. The PBC monopolized banking functions, and acted as the central bank and a commercial bank. Enterprises were required to hold accounts in the PBC, which enabled it to monitor their financial positions. In addition, the People's Construction Bank of China handled state budget allocation of capital construction investment and financial control of construction enterprises. The interest rate was controlled at a low level by the government and changed little from year to year. Decentralization and local intervention in the financial sector were at a minimum.

China's financial system has experienced dramatic reforms since the late 1970s. First, numerous bank and nonbank financial institutions have been established. In 1978, the PBC was separated from the Ministry of Finance, and the state decided in 1983 to transform the PBC into a central bank. Four major commercial banks were established or became independent, including the Bank of China (specializing in foreign exchange), the Agricultural Bank of China, the Industrial and Commercial Bank of China, and the People's Construction Bank of China. Two more banks, the China International Trust and Investment Corporation (CITIC) and the China Investment Bank, were later added to the list. With the launch of comprehensive reforms in the mid-1980s, more financial institutions were established in the following years, including the Bank of Communications, the Import-Export Bank, the Long-Term Development and Credit Bank, and regional and specialized banks. Urban and rural credit cooperatives and trust and investment companies have also grown rapidly. By 1996, there were 16,101 institutions in the banking and insurance sectors that employed 2.1 million workers.

Secondly, financial markets and instruments have been reopened. Bonds were issued as early as 1981, and have become an important source for financing infrastructure investment projects. Stock markets have also been established, although much later, and they are more volatile. The Shanghai Securities Exchange and the Shenzhen Stock Exchange were officially opened in December 1990 and July 1991, respectively. Shares are available to foreign investors (B shares). While the stock markets are still at an early stage of development, they have stimulated the growth of shareholding companies in China.

Thirdly, measures have been taken to improve financial management. These include separating the central banking function from commercial banking, decentralizing bank management and turning banks into independent enterprises. Direct budget grants to enterprises have been replaced with interest-bearing loans to improve efficiency. Interest rates are partially floating to attract savings and to adjust to the emerging market economy. In 1979, the deposit rate for individual one-year fixed deposits was 3.96 percent, which was raised to 11.34 percent in 1990 but declined to 10.98 percent in 1996, in response to inflation and financial reforms. Reforms have improved banking specialization and have gradually introduced competition among banks. The scope of banking has been enlarged to serve different demands.

Lastly, China has also opened its door to international financial markets and quickly developed its external bank networks. China's legitimate representation in the World Bank and IMF was established in 1980. The China Investment Bank was created in 1981 to disburse international financial institutions and to finance foreign-invested projects and technology imports. In the early 1980s, foreign banks were allowed to set up representative offices in some coastal cities. In 1985, they were permitted to operate their branches in China, and in 1990, Shanghai was opened to foreign banks. In 1992, about 47 foreign-funded banking and financial institutions and 218 representative offices of foreign banks were set up in China (Chai 1997). By the end of 1995, there were more than 100 branches of foreign banks operating in China. The scope of international banking activity expanded rapidly during the mid-1990s, although the establishment and operation of foreign banks in China are still heavily regulated.

As a result of those reforms, the role of the banking system in economic development has become more important, and China has emerged as one of the world's largest banking markets. The emergence of the banking system has contributed to the declining control of the state budget in China. Sizes of saving and loans have been increasing rapidly, which has provided more resources to enterprises and individuals. As I will analyze in the following sections, more funds have been channeled into more profitable sectors.

However, Chinese banking sector and monetary policy are still troubled by many problems. The extent of financial reforms is still limited and has not significantly improved investment efficiency. The banking system is still under strict state control, and the level of marketization and privatization remains very limited (Ma 1996). The PBC is still less effective in adjusting interest rates, and poor lending decisions led to mounting debts for many banks and companies. Banks have channeled a large share of loans to inefficient, money-losing SOEs (Lardy 1998), which has become the biggest problem facing China's banks. Local governments are heavily influencing the lending practice of local branches of banks, often in order to protect the interests of regions and at the sacrifice of efficiency. New mechanisms of regulation enforcement are still not functioning well or have yet to be developed.

In addition, the process of financial reform was uneven among regions. First, like other reforms, financial reforms tend to launch first in the coastal region and rich areas. Banks in these regions also tend to enjoy more open and decentralized policies. Secondly, regional banks tend to be established first in more developed coastal provinces. For example, Shenzhen Development Bank (in Guangdong province) was the first one to list its shares on the stock exchange. Minsheng Bank is the first private bank established in China. Other major regional banks include Guangdong Development Bank, the Bank of China Merchants (in Shekou, Guangdong province), Fujian Industrial Bank, and Pudong Development Bank, all located in the coastal region. Thirdly, foreign exchange markets are opened only in Shanghai and Shenzhen, again in coastal China. A large percentage of the enterprises listed is from the coastal provinces. Lastly, coastal cities are the first ones where international financial institutions are allowed to operate. Financial reforms have provided regions new power to obtain financial resources that are needed for economic growth, but the coastal provinces tend to obtain more favorable reform policies.

Financial reforms have dramatically increased savings available for investment, especially in the coastal provinces. The coastal region has per capita bank deposits, loans, and savings that are double those in the central and western regions. Financial reforms have also partially forced banks to provide more loans to more profitable enterprises, such as FIEs and TVEs, which contributed to the development of non-state sectors, although SOEs, especially those in the interior region, still dominate bank loans. While interregional transfer of funds among banks is limited by the central government, such transfers do take place among regions.

3.2 Revenue, expenditure, and interregional transfers

3.2.1 Regional patterns of revenue and expenditure

Overall, China's revenue and expenditure increased in the 1950s, stagnated from the 1960s to the mid-1970s, but have grown dramatically since economic reforms. The last four decades recorded four periods of decline of budget: 1961–62, 1967–68, 1974–76 and 1979–80, corresponding to political and economic crises. In terms of revenue sources, the share of revenues from industry grew rapidly in 1950–83 but has since staggered. The share of revenue from agriculture was high in the 1950s but has gradually declined. The commerce sector stagnated in Mao's era but its development has been accelerated. In terms of ownership structure, the share of revenue generated by SOEs has declined steadily since 1978, reflecting the inability of the state to revitalize SOEs.

Budgetary expenditure increased rapidly in the 1950s, stagnated from the 1960s to the mid-1970s, and has reemerged since the late 1970s. It experienced four troughs: 1961–62, 1967–68, 1974–76 and 1979–80, corresponding to revenue troughs and economic retrenchment. Major budgetary expenditures financed economic development, especially capital construction investment in SOEs. The share of agricultural spending was low in China, particularly the spending from central government. The share of spending in national defense increased dramatically in 1964–71 when the defense-oriented Third Front project was implemented. The most rapidly increasing expenditures in post-Mao China were education, administration and price subsidies.

A comparison of revenue and expenditure shows that deficits under Mao took place in the late 1950s, the early 1960s, 1967, and 1974–76. In the post-Mao period, however, only 1978 and 1985 registered a budget surplus. Budget deficit was mainly a problem associated with the central government, as the center required the regions to balance their budgets. Solutions of the center included increasing money supply, extracting loans from the provinces, issuing treasury bonds, borrowing from international organizations, and shifting more budgetary responsibilities to the provinces. However, these efforts did not significantly reduce the budget deficit as the center had not fundamentally reduced its spending. The central government retained strong control over budget funds, as evidenced by the fact that the center still receives and spends 40 percent of the budgetary funds.

Regions differ considerably in revenue growth. Table 3.4 shows that during Mao's era coastal provinces, especially Shanghai, Beijing, Liaoning, Hebei, and Tianjin, registered more rapid revenue growth than most of the central and western provinces, although Qinghai, Shaanxi, and Shanxi also experienced rapid growth. The provinces with the slowest growth rates included Neimenggu, Jilin, Jiangxi, and Guizhou. From 1978 to 1990, the eastern region showed dramatic disparities: the growth rates of Shanghai, Beijing, Tianjin, and Liaoning were among the lowest in the nation while those of Zhejiang, Guangdong, and Fujian were among the highest. Rapid growth was registered in the western region. From 1990 to 1995, the revenue condition in the three municipalities and traditional industrial bases were improved, although growth rates of Guangdong, Fujian and Shandong were still higher. Revenue growth was the slowest in the western region.

Regional patterns of expenditure are somewhat different from that of revenue (Table 3.5). During Mao's era, the western region experienced the fastest growth in per capita expenditure although its revenue growth was slower than the eastern region, suggesting the existence of large interregional transfers to the backward regions. Since reforms, however, the eastern region has consistently outspent the central and western regions. During the 1978–90 period, the eastern provinces of Guangdong, Zhejiang, Fujian, Liaoning, Jiangsu, and Shandong experienced faster growth while Shanghai, Beijing, Tianjin, Guangxi, and Hebei were among the lowest in the nation.

Table 3.4 Regional variations of per capita revenue in selected provinces (yuan), 1952–95

	1952	*1978*	*1990*	*1995*	*1952–78 percent change*	*1978–90 percent change*	*1990–95 percent change*
Average	17	100	173	248	478	73	43
Eastern region	22	158	229	358	631	45	56
Beijing	84	594	702	919	607	18	31
Tianjin	88	542	508	658	513	−6	30
Hebei	14	89	132	186	557	48	41
Liaoning	40	282	326	450	597	15	38
Shanghai	45	1541	1180	1548	3363	−23	31
Jiangsu	18	105	201	245	476	92	22
Zhejiang	18	73	244	271	317	233	11
Fujian	17	62	188	361	253	205	92
Shandong	16	80	128	206	404	61	61
Guangdong	23	73	197	541	218	170	175
Guangxi	11	42	112	174	273	166	55
Central region	16	68	134	177	331	98	32
Shanxi	13	81	185	234	518	128	26
Neimenggu	19	38	152	193	104	302	27
Jilin	34	76	204	243	128	167	19
Heilongjiang	43	202	216	273	375	7	26
Jiangxi	14	38	107	158	178	179	48
Western region	12	49	134	162	312	175	21
Guizhou	8	23	110	111	185	372	1
Yunnan	11	38	208	246	245	447	18
Shaanxi	12	71	134	145	500	89	8
Qinghai	6	79	162	187	1179	104	15

Source: SSB 1990; 1992; 1996.
Note: Guangdong includes Hainan.

Dramatic disparities also exist in the central and western regions as some provinces, such as Yunnan and Sichuan, experienced rapid growth in expenditure while others, such as Qinghai and Ningxia, registered growth rates far below the national average. From 1990 to 1995, the gap in expenditure between the coastal and western regions rose further. Shanghai led the nation in expenditure growth, much faster than its revenue, while several western provinces, including Ningxia, Qinghai, Guizhou, and Gansu, had expenditure growth far below the national average.

The percentage of revenue received directly by the central government increased while that of the provinces declined. In 1978, only 14.7 percent of revenue was directly collected by the center and the figure for 1991 was 38.8 percent. The key to deficit was that the central government did not fundamentally change its spending behavior. It still actively initiated a large

Table 3.5 Regional variations of per capita expenditure in selected provinces (yuan), 1952–95

	1952	*1978*	*1990*	*1995*	*1952–78 percent change*	*1978–90 percent change*	*1990–95 percent change*
Average	8	63	182	402	668	188	121
Eastern region	10	68	212	533	607	211	151
Beijing	41	240	633	1234	489	164	95
Tianjin	25	200	455	991	715	127	118
Hebei	6	64	142	297	1016	121	110
Liaoning	20	92	308	669	362	234	117
Shanghai	37	237	566	1837	546	139	225
Jiangsu	6	49	149	359	762	207	140
Zhejiang	5	46	192	417	835	314	117
Fujian	10	62	225	530	522	265	135
Shandong	7	44	146	317	568	230	117
Guangdong	11	51	240	748	361	367	212
Guangxi	9	61	154	309	602	153	100
Central region	8	59	145	331	653	145	127
Neimenggu	14	103	282	447	613	175	59
Jilin	10	76	289	466	694	279	62
Heilongjiang	15	101	262	472	591	160	80
Western region	6	60	184	349	883	207	90
Sichuan	5	37	132	245	572	259	86
Guizhou	4	46	149	243	950	225	64
Yunnan	5	59	243	589	1014	311	142
Gansu	6	77	204	334	1176	166	64
Qinghai	11	186	382	599	1566	105	57
Ningxia	13	162	318	448	1113	96	41

Source: SSB 1990; 1992; 1996.
Note: Guangdong includes Hainan.

amount of projects and increased administrative spending. The share of expenditure between the center and provinces only changed slightly. In 1978, 46.9 percent of expenditure was through the center, and in 1992, 41.4 percent. The central government still retains strong control over budget funds. During the early 1990s, it received and spent 40 percent of the budgetary funds and managed 40 percent of the extrabudgetary funds.

A rapid increase of public sector extrabudgetary funds is one of the most dramatic features of China's fiscal reform (Fig. 3.1). This phenomenon is astonishingly similar to Eastern European transition economies where a striking feature of local government finances was the proliferation of extrabudgetary funds (Bird and Wallich 1994). In 1991, extrabudgetary revenue and expenditure reached 324.3 and 309.2 billion yuan, respectively, almost matching budgetary revenue (361.1 billion yuan) and expenditure (381.4

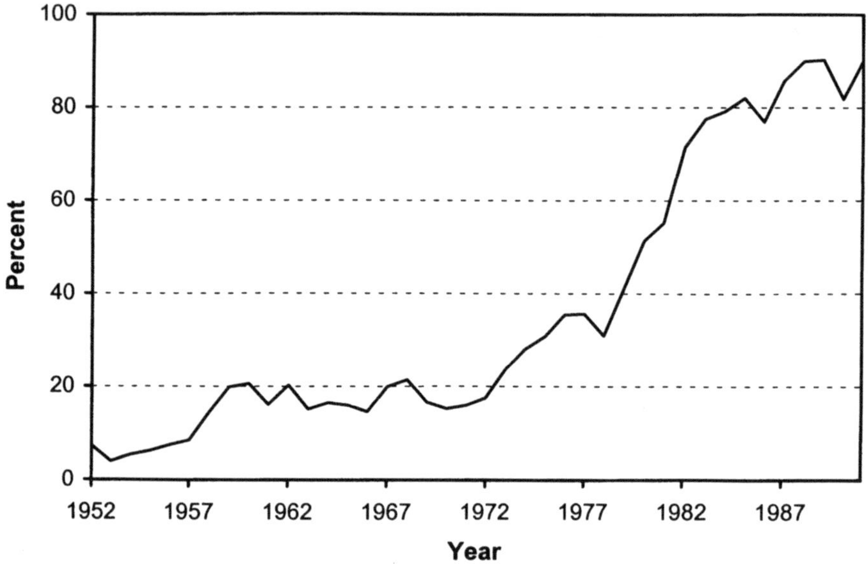

Figure 3.1 Increasing extrabudgetary revenue as a percentage of budgetary revenue, 1952–91.
Source: MOF 1992.

billion yuan). Extrabudgetary funds are not formally included in the official budget, and are less strictly controlled by the central government. Major sources of extrabudgetary revenue included revenue received from SOEs that had been managed by regions, fees collected for government services, and income managed by local government bureaus. In 1991, they were composed of 76.4 percent, 21.5 percent and 2.1 percent of total extrabudgetary revenue, respectively.

In such coastal provinces as Shanghai, Beijing, Jiangsu, and Liaoning, extrabudgetary expenditures surpassed their budgetary ones while those in interior provinces were considerably less than budgetary funds (Table 3.6). The rapid growth of extrabudgetary funds provided regions, particularly coastal provinces, more expenditure for economic development. The growth of extrabudgetary funds further reduced the equalization function of public finance. While the central government increased the control of extrabudgetary funds, the local governments created new sources of financing, such as fines, administrative charges, and forced contributions by enterprises.

3.2.2 Interregional resource transfers

Interregional resource transfer is often used in socialist countries to implement industrialization and regional development policies. The extent of transfer, however, is difficult to measure, as such transfer is often ad hoc

Table 3.6 Provincial variations in revenues and expenditures (billion yuan), 1990

	Budgetary			*Extrabudgetary*		
	Revenue	*Expenditure*	*Expenditure as percent of revenue*	*Revenue*	*Expenditure*	*Expenditure as percent of revenue*
Eastern region	119.60	111.28	93.0	91.58	92.97	101.5
Shanghai	15.77	7.57	48.0	10.38	10.74	103.5
Beijing	7.62	6.87	90.2	8.59	7.81	90.8
Tianjin	12.93	12.22	94.6	3.91	3.77	96.4
Liaoning	16.15	15.15	93.8	13.32	15.73	118.1
Hebei	8.12	8.73	107.6	7.66	7.74	101.0
Shandong	10.91	12.39	113.5	9.77	10.04	102.8
Jiangsu	13.63	10.10	74.1	11.53	11.43	99.2
Zhejiang	10.16	8.02	79.0	7.70	7.59	98.6
Fujian	5.71	6.85	120.0	4.53	4.13	91.0
Guangdong	13.10	15.07	115.0	9.84	9.20	93.5
Hainan	0.74	1.74	235.7	0.55	0.56	102.4
Guangxi	4.76	6.58	138.1	3.83	4.24	110.8
Central region	56.25	66.83	118.8	45.95	47.76	103.9
Shanxi	5.38	5.69	105.9	5.19	5.12	98.6
Neimenggu	3.30	6.09	184.7	3.76	3.92	104.2
Jilin	5.07	7.17	141.4	5.74	5.78	100.8
Heilongjiang	7.66	9.27	121.1	5.92	6.31	106.6
Anhui	5.45	6.16	113.0	4.16	4.65	111.8
Jiangxi	4.06	5.08	125.0	3.61	3.67	101.8
Henan	8.60	9.19	106.9	5.85	6.09	104.0
Hubei	9.51	9.95	104.7	5.47	5.89	107.6
Hunan	7.24	8.24	113.8	6.26	6.33	101.2
Western region	34.79	47.79	137.4	26.00	26.22	100.8
Sichuan	11.98	14.27	119.1	9.51	9.96	104.7
Guizhou	3.61	4.86	134.6	2.06	1.97	95.5
Yunnan	7.74	9.08	117.2	4.46	4.30	96.3
Xizang	0.02	1.29	7177.8	0.20	0.20	102.0
Shaanxi	4.44	5.71	128.6	3.81	3.93	103.2
Gansu	3.42	4.59	134.3	2.34	2.23	95.2
Qinghai	0.72	1.71	236.6	0.87	0.79	91.1
Ningxia	0.67	1.50	222.3	0.71	0.70	97.9
Xingjiang	2.18	4.78	219.6	2.05	2.15	105.1

Source: MOF 1992.

and there is also a lack of data. A comparison of central and local shares of budget can indicate the resources available for the central government to spend, and a comparison of local revenue with expenditure can show the extent of interregional budget transfer. As regions had to follow non-deficit financing, those collecting more revenue than expenditure tended to hand

Table 3.7 Central and provincial revenues and expenditures (billion yuan), 1957–95

	Revenue					*Expediture*				
Year	*Total*	*Center*	*Percent*	*Province*	*Percent*	*Total*	*Center*	*Percent*	*Province*	*Percent*
1957	31	14	45	17	55	30	22	72	9	28
1962	31	9	30	22	70	31	19	63	11	37
1965	47	16	33	32	67	47	29	62	18	38
1970	66	18	28	48	72	65	38	59	27	41
1975	82	10	12	72	88	82	41	50	41	50
1980	116	28	24	88	76	123	67	54	56	46
1985	200	77	39	124	62	200	80	40	121	61
1990	294	99	34	194	66	308	100	32	208	68
1995	624	326	52	299	48	682	200	29	483	71

Source: SSB 1993; 1996.

over the surplus to the central government while those with less revenue tended to receive transfers.

A comparison of revenue with expenditure in both central and provincial budgets is helpful in determining central control over budget resource and its implications for interregional transfer. As shown in Table 3.7, shares of central and provincial governments over budget resources changed over time. The center controlled much of the resources during the 1950s, indicating the central government had more resources under control. As I will show in the following sections, these resources were used to finance China's massive industrialization during the First FYP period. But following the change of central–provincial fiscal arrangement, the central share of budget declined greatly during the 1970s, indicating the increasing resources controlled by local governments, which partially contributed to the development of rural enterprises in China.

The level of central control declined somewhat in the 1980s, but through the implementation of the 1994 reform, the central government has regained its control over budget resources. In 1980, only 14.7 percent of revenue was directly collected by the center, which increased to 24 percent in 1990. The share of expenditure between the center and provinces has also changed. In 1980, 54 percent of expenditure was through the center, which declined to 29 percent in 1995. As a result of the implementation of a new fiscal system in 1994, central government increased its control over budget resources. It still retains some control over budget funds. A comparison of revenue and expenditure shows that during much of the reform period the central government was in deficit. The deficit situation in the provinces became commonplace only in the 1990s as, during the early years of reforms, the provinces had to follow non-deficit financing.

A comparison of revenue and expenditure, to a certain extent, shows the level of interregional fiscal transfer.[5] As the regions had to follow non-deficit financing, those collecting more revenue than expenditure tended to hand

Table 3.8 Ratio of expenditure to revenue in selected provinces, 1952–91

	1952	*1957*	*1962*	*1965*	*1970*	*1975*	*1978*	*1985*	*1991*
Average	47	56	48	55	58	58	63	89	104
Eastern region	45	47	36	39	37	41	45	66	91
Shanghai	82	81	9	12	13	19	15	25	53
Beijing	49	35	28	40	31	41	40	63	90
Tianjin	28	20	25	26	23	27	37	56	82
Liaoning	49	26	37	41	35	35	33	67	94
Shandong	42	46	51	58	46	46	55	76	99
Jiangsu	31	48	47	41	40	41	46	58	89
Zhejiang	28	40	41	37	47	75	63	64	80
Guangdong	49	48	48	48	52	52	70	97	100
Fujian	57	77	71	76	129	103	100	122	112
Central region	50	60	72	78	95	82	87	121	118
Jilin	29	38	81	74	94	97	100	159	127
Heilongjiang	34	52	74	80	107	40	50	119	116
Western region	51	77	73	98	117	120	123	156	128
Guizhou	53	74	79	109	161	479	196	162	122
Qinghai	180	261	229	176	226	220	234	420	207

over the surplus to the central government while those with less revenue tended to receive transfers.[6] Eastern provinces had lower ratios than western provinces, especially during the 1960s and early 1970s, which indicated the existence of interregional budget transfer (Table 3.8). Regional disparities between revenue and expenditure declined in post-Mao China, indicating the reduction of interregional transfer and a closer correspondence between revenue and expenditure. The decline of budgetary transfers has led to an increase of spending in rich counties in the 1980s, while some poor counties experienced budget distress. In 1991, three groups of provinces can be classified by comparing local revenue with expenditure:

1. Revenue-surplus and contributing provinces: the eastern provinces of Shanghai, Beijing, Tianjin, Jiangsu, Zhejiang, and Liaoning had a ratio of expenditure to revenue of less than 100, indicating that they contributed to the central government coffers.
2. Zero-balance provinces, such as Hebei, Henan, Hubei, and Shanxi.
3. Revenue-deficit and subsidized provinces: largely minority and poor interior provinces.

The coastal provinces of Fujian and Hainan also received subsidies. The ratios of extrabudgetary revenue to extrabudgetary expenditure differ only slightly among provinces, indicating that much less extrabudgetary revenues were transferred interregionally.

Interregional resource transfer is an extremely complicated issue, especially when the transfer is irregular. Wong (1997), for example, listed 20 items

Table 3.9 Rate of interregional national income transfer in selected provinces, 1953–92

	1953–57	*1958–62*	*1963–65*	*1966–70*	*1971–75*	*1976–80*	*1981–85*	*1986–92*
Eastern region	−9.4	−15.6	−15.7	−20.2	−21.2	−20.5	−12.8	−4
Beijing	68.2	7.8	−5.9	−14.7	−23.4	−26.4	−7.7	11.6
Liaoning	−17.7	−29.1	−11.4	−24.2	−30.3	−26.6	−18.1	−5.8
Shanghai	−41.9	−51.1	−55.2	−58.6	56.2	−60.7	−43	−14.3
Jiangsu	−10.9	−7.6	−11.3	−15	−14.8	−14.2	−13.5	−7.1
Zhejiang	−9.3	−10.6	−16.1	—	—	−8.6	−8.5	−4.7
Guangdong	3	−1.2	−6	−4.3	−3.5	−8.3	−0.3	−4.5
Central region	−1.4	−1.6	−4.2	−6.2	−4.1	−2.9	−1.2	−1
Jilin	2.1	−12.5	−7.4	−2.8	−3.2	6.9	7.4	6.8
Heilongjiang	−67	−15.2	−9.9	−18.4	−12.9	−20	−10.4	−1.6
Henan	3	12.8	15.9	8.4	−1.9	−0.8	−2.9	−3.1
Western region	7.6	8.6	9.1	20.9	15.2	10.8	12.2	11.1
Guizhou	8.9	5.3	29.7	38.8	46.7	24.6	14.4	6.4
Yunnan	3	13.8	7	27	17.7	22.9	17.1	7
Ningxia	12.7	39.2	40.7	36.7	16	25.8	31.4	31

Source: Based on H. Wei *et al.* 1997.

under central earmarked grants in 1993. On the one hand, the central government transfers resources to poor regions through quota subsidies, grants, earmarked subsidies, and other welfare policies. On the other hand, the price structure tends to transfer hidden resources to the rich regions, as agricultural products and raw materials tend to be priced lower (the scissor's gap).[7] Scholars have compared national income to show the extent of interregional transfers (e.g., Zhou 1996; Wei *et al.* 1997). Resources tend to move out of the eastern region to the western region, but the extent of transfer declined after the early 1980s (Table 3.9).[8] Zhou (1996) argues that much of the resources was used by the central government rather than by the poor regions. The decline of interregional transfer during the reform period caused much concern.

3.3 Investment and investment systems

3.3.1 Investment and regional development

Investment and its contribution to regional development have long been a subject of theoretical work, and dominant theories of regional development assume free geographic mobility of capital. Neoclassical regional growth theories maintain that efficient markets and factor mobility tend to equalize regional differentials (e.g., Borts and Stein 1964; Williamson 1965; McCombie 1988). Divergence and structural theories, particularly those with Marxist political economy perspectives, also build on the notion of free geographic mobility of capital, although they argue that backwash effects and capital accumulation tend to maintain and reinforce regional inequality (e.g., Myrdal 1957; Soja 1980; Smith 1984).

These bodies of literature, however, have three major limitations. First, the impact of state policy on investment patterns has rarely been studied. While the role of the state has captured scholarly attentions (e.g., Appelbaum and Henderson 1992), more studies are needed to unfold the content of investment policy and the role of the state in investment allocation. Secondly, dominant theories of regional development have overemphasized the free mobility of capital in determining regional development. Thirdly, capital flow across countries has increased rapidly with the globalization of economic activities, and foreign investment has emerged as an important part of total investment. However, existing studies on regional investment tend to exclude foreign investment, while the literature on the latter only reveals a partial picture of investment. Therefore, more work is needed to improve the understanding of capital investment and regional development, especially in the context of transition economies.

In the following sections, I examine emerging patterns and underlying factors of regional investment in China, hoping to address some of the gaps in the literature. I examine particularly fixed assets investment (or investment in fixed assets) (*guding zichan touzi*), which includes capital construction investment (CCI) and technical renovation investment. Regional allocation of fixed investment (especially CCISOEs) is considered a key instrument in China's industrialization and regional development policy (Zhou 1993; Ma and Wei 1997). Based on a review of the literature and my empirical investigations, I have generated several arguments. First, investment distribution is geographically uneven. During the last decade, empirical studies have shown that investment distribution is uneven and is influenced by previous investment patterns, technical innovation, and agglomeration economies (Clark *et al.* 1986). Research on venture capital (VC), in particular, suggests that VC is geographically concentrated in established financial centers and in centers of high-technology industry (e.g., Martin 1989; Florida and Smith 1993).[9] This unevenness is a central feature of regional investment, which needs to constantly monitor the change of investment patterns and to examine factors underlying the unevenness.

Secondly, the state and economic reforms in transition economies have had a profound impact on investment patterns. Since the late 1970s China has dramatically restructured its political economy and space. Researchers have shown the declining control of the state over revenue and investment (Naughton 1987; Wei 1996), which has prompted concerned scholars and government officials to call for recentralization and strengthening of state capacity. While the control and capacity of the state have declined in China, the state is still actively initiating new policies for economic development and has recently recentralized its control over the economy. Moreover, state policy in China is extremely complex and changing constantly. An analysis of reform policies and process, including investment reform, is essential for the understanding of structural and spatial changes of investment and regional development in China.

Thirdly, regions and global forces are playing increasing roles in shaping China's uneven patterns of regional investment and development. Economic reforms, particularly economic decentralization, have empowered local states that act like development and entrepreneur states to promote local development. Regions have emerged as important factors and agents in capital accumulation and investment allocation (Huang 1996), contributing significantly to regional disparities in investment. Their abilities to adapt to change and to capture the opportunities created by economic reforms are at least partially related to previous development experiences and investment patterns. In addition, global forces directly affect investment, and should be examined in order to understand investment patterns. Therefore, I adopt the framework that the nexus of global, state, and regional forces determines investment patterns and regional development.

Lastly, fixed investment is largely accumulated within regions, and investment flows across regions are still very limited in China. Dominant theories of investment and regional development are based on the assumption of free mobility of capital (e.g., Borts and Stein 1964; Williamson 1965; Soja 1980; Smith 1984). However, an emerging body of research has argued for market imperfections and spatial rigidities, and the limitation to capital mobility in developed countries (e.g., Clark *et al.* 1986; Gertler 1986; Martin 1989; Florida and Smith 1993). In contrast to the notions of market perfection and free mobility of capital, these studies tend to emphasize disequilibrium, path dependence, and increasing returns to scale. Developing countries such as China are highly segmented by regionalism and the poor transportation, and the interregional flow of capital is limited or even prohibited by state policies.

The recent release of statistics on investment has made my attempt to study regional investment in China possible (SSB 1986–97, 1997c). SSB (1997c), in particular, has provided detailed temporal data on fixed investment in China from 1950 to 1995. In addition, 'white' reports of the government on investment have provided useful information on China's investment reforms and forces underlying investment patterns (e.g., SPC 1996–97). My informal interviews with Chinese officials and scholars also help in the understanding of the complexity of investment in a transition economy and in the interpretation of changing investment patterns.

3.3.2 China's investment systems

China established its investment system in the early 1950s to take control over the economy and to promote the nation's industrial and spatial development. This investment system was highly centralized through centralized planning, centralized administration and state-controlled banks. The central government largely controlled the size, structure, and location of fixed investment, and gave little consideration to investment efficiency and local autonomy. But occasionally, the level of central control was reduced somewhat, such as during the Great Leap Forward and the early 1970s when Mao encouraged decentralization and rural development. During Mao's era, state budgetary investment often accounted for 80 percent of CCISOE. Investment was concentrated in heavy industry and SOEs, under the influence of the Soviet model of socialist industrialization.

This centralized investment system partially enabled the government to mobilize its resources among sectors and regions for the purposes of industrialization, national defense, and balanced development. This system, however, had several problems, most of which were typical to Soviet-type command economies (Kornai 1992):

1 overcentralized investment planning and management that gave little incentive to local governments, enterprises, and individuals;
2 overconcentration of investment in heavy industry and the defense sector, and little investment in infrastructure and services;
3 overaccumulation of capital at the ignorance of consumption;
4 low investment efficiency;
5 the lack of project feasibility studies, and chaos in the construction process (Peng *et al.* 1989; Xia and Li 1993).[10]

These problems, as well as changing international and domestic conditions, prompted post-Mao leaders to reform Mao's investment systems.

In the late 1970s, the Chinese government introduced investment reforms to solve problems in investment allocation and management. Policies of adjustment were implemented in the following years to adjust size, structure, and spatial distribution of fixed investment. With the focus of reforms shifting to urban reforms in the mid-1980s, more investment policies were introduced, focusing on investment decentralization, investment structural adjustment, and project management. A significant change was decreed, from a system consisting of profit remittances from the enterprises to the government and targeted budgetary grants for investments from government to enterprises, to a system which involves profit retention, enterprise taxes, and the financing of investment through bank loans (*buo gai dai*) (Yenal 1990).[11] In 1988, the government established capital construction investment funds and specialized investment corporations to adjust investment allocation and to provide more funds for key-point construction.[12] Indeed, since the late 1970s, a series of measures have been introduced to improve investment allocation and management, including the requirement for project evaluation, the introduction of the investment orientation tax, and the establishment of state-controlled specialized investment banks and companies. Deeper reforms were implemented in the 1990s to stimulate the development of market economies and to further improve investment allocation and management.

Investment decentralization has provided local governments with considerable power in capital accumulation and investment allocation. As shown in Fig. 3.2, fixed investment financed through the state budget has declined dramatically (only 3.1 percent in 1995). China's investment reforms have decentralized most of the investment allocation and management functions to the local governments and enterprises. Cities and regions are allowed to keep increasing portions of revenue and income for local investment. Local governments, such as Guangdong, Fujian and Jiangsu, have used their increased power to expand revenue bases and extrabudgetary funds. Local authorities and enterprises have retained greater shares of production profits and depreciation funds and buttressed their financial position through more bank loans and increased extrabudgetary income.

The project approval process, through which the central government controlled investment activities, has been largely decentralized to local governments, while the central government focuses on large-scale projects and priority sectors such as energy and transportation. During the 1980s, investment projects that required approval by the State Planning Commission were raised from 10 to 30 million yuan, and that by the State Council from 100 to 200 million yuan, which were further raised in the 1990s. From 1984 to 1994, the percentage of CCISOE projects (243.1 billion yuan in 1994) administered by local governments increased from 40.6 percent to 59.8 percent (SSB 1997c). In addition, much of the investment in the projects administered

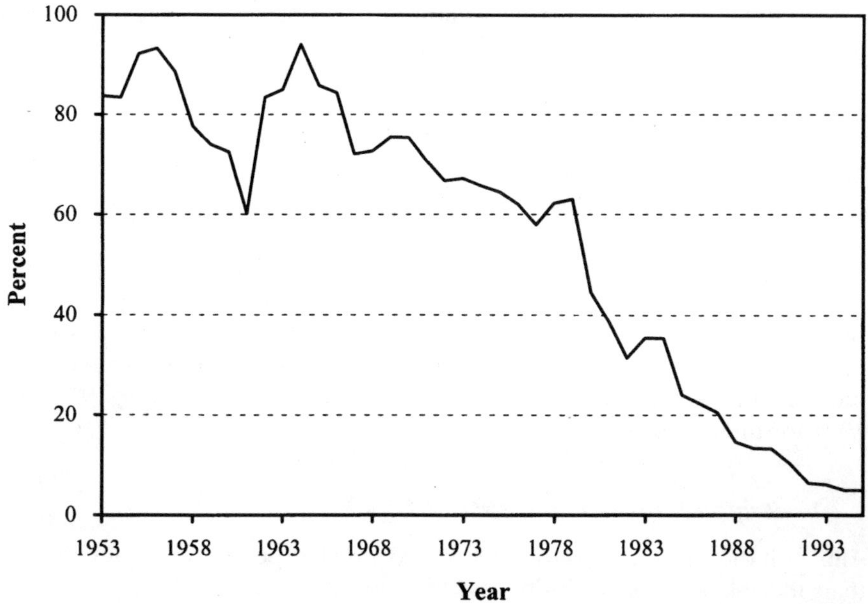

Figure 3.2 Declining budgetary investment as a percentage of fixed investment, 1953–95.
Source: SSB 1997c.

by the central government is funded by bank loans, rather than directly by the state budget. Local investment is often required to match the investment provided by the central state. With an increasing proportion of output being produced by non-state enterprises, the control of the central government has declined further as non-state sector investment, especially private investment, is largely under local control.

However, investment decentralization has not dramatically improved central–local relations and macroeconomic management. Central–local conflicts have persisted, as the central and local governments have to negotiate over profit remittances, taxes, investment grants, project allocation, and many other areas. The central government often has to rely on recentralization of financial resources and administrative power to implement central policy due to the lack of macro adjustment mechanisms. Localities have used a variety of strategies, such as breaking projects down into separate small subprojects and underestimating project costs, to avoid the project approval system which gives the central government the power to impose its preferences on investment allocation (Singh 1992). Local strategies and rent seeking have prompted the construction of numerous duplicated projects, which are often against the policy of the central government but are largely protected by local governments. As I will elaborate in the following sections, local efforts have significantly contributed to the rapid increase of fixed investment and have made the attempt of the central government to adjust investment size and allocation difficult.

3.4 Regional investment under Mao

To shed some light on the regional patterns of investment, I examined patterns and underlying factors of fixed investment in the three regions and the provinces. Following Ma and Wei (1997), I divide the history of investment from 1950 to 1996 into four periods: the first period, 1950–64, covered the First Five-Year Plan and the Great Leap Forward and its aftermath; the second, 1965–71, was the Third Front period; the third period, 1972–77, marked the beginning of China's economic adjustment; full-scale reforms were implemented during the reform period. I have also selected 1953, 1965, 1978, and 1995 for more detailed analysis.

3.4.1 Regional investment, 1950–64

Although the eastern region was allocated a higher percentage of investment than the inland regions during the first decade of the PRC, the backward inland regions received serious attention from the state for interior development (Fig. 3.3). China's central planners showed their concern for spatial equality in development during the First Five-Year Plan (1953–57) (Li *et al.* 1990; Lu *et al.* 1990; Peng *et al.* 1989). In response to the need for a quick economic recovery after the civil war, the state invested more than half (51 percent) of the CCISOE in the coastal region to rehabilitate the old industrial bases in east and northeast China that had been damaged by the war. The locations of many of the development projects found in the inland provinces were determined by a number of factors, including proximity of raw materials, defense considerations, national and local economic needs, minimization of transportation costs, regional self-sufficiency, and a spatially more balanced distribution of the productive forces (Lu *et al.* 1990; Ma and Wei 1997).

In 1953, investment preference was given to the provinces where there was a need for industrial rehabilitation and improvement, and to interior provinces where new construction was soon underway (Fig. 3.4). Beijing, the nation's capital, led the country by a large margin, and since then it has remained a leading receiver of fixed investment. Considerable investment went to the three northeast provinces of Liaoning, Jilin, and Heilongjiang, which had developed a strong industrial foundation. Per capita CCISOE was also high in Shanghai and Tianjin, China's leading commercial and industrial centers. Beijing, Shanghai, Tianjin, and the three northeastern industrial provinces accounted for 47.6 percent of the total CCISOE. Among the provinces that received the smallest amounts of total and per capita CCISOE were Zhejiang, Fujian, and Guangxi along the coast and Guizhou in the western

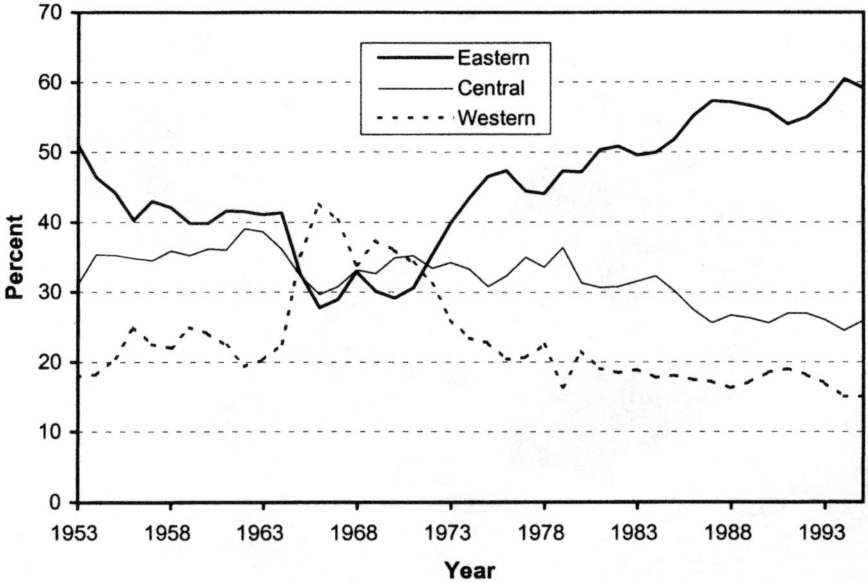

Figure 3.3 Regional distribution of capital construction investment in state-owned enterprises, 1953–95.

Source: SSB 1990–96. Ma and Wei 1997. Determinants of state investment in China, *Journal of Economic and Social Geography* 88(3): Fig. 4. Reproduced with permission of Blackwell Publishers Ltd, Oxford.

region. The remote border provinces of Tibet, Xinjiang, and Qinghai also had higher per capita CCISOE, partially due to these provinces' sparse populations. During the First FYP, the inland regions received 55.9 percent of CCISOE, of which over 60 percent was used for new construction, whereas in the eastern region only 25 percent of the CCISOE was earmarked for new construction and 75 percent for the expansion and renovation of existing projects (Peng *et al.* 1989).

The locations of the First FYP's 156 key-point projects imported from the former Soviet Union can further show regional policy of the central government. These projects cost the state a total of 20.3 billion yuan to build (SSB 1991b). Of the 150 projects actually constructed, about two-thirds were allocated in the central part of China, while 12 provinces, eight of them coastal in location and four along the border, received no projects at all. Investment in the 98 civilian key-point projects amounted to 14.3 billion yuan, of which 72 projects with 68 percent of the investment went to the inland provinces (Table 3.10). The central region of China received more than half of the total number of projects as well as more than half the total amount of investment. Of the eight provinces with over 1 billion yuan of investment, only

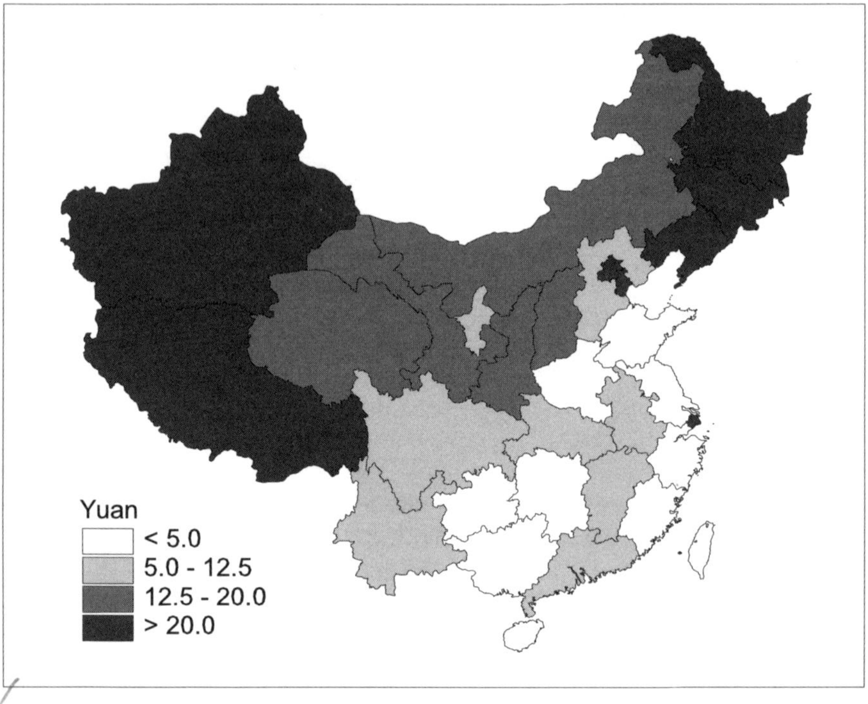

Figure 3.4 Per capita capital construction investment in state-owned enterprises, 1953.
Source: SSB 1990.

Liaoning is in the eastern region although the investment there was the largest (mainly for the steel complex in Anshan), and eight of the 11 coastal provinces received no key-point project at all.

Such a pattern of investment was a consequence of several factors, including the need to rehabilitate the existing national economic centers located along the coast, defense considerations, and the desire to alter the uneven spatial pattern of development. During this period, China faced two incompatible goals of national development: the goal of economic efficiency that demanded spatial concentration of resources; and the goal of reducing regional gaps of development that required resource dispersal. In addition, threats to China's security were perceived by the Chinese leaders to have existed along China's pacific coast, due to the involvement of the American forces there to safeguard peace in the Korean peninsula. Tension also existed in the Taiwan Strait as Taiwan was seen as a renegade province of China to be liberated while the leadership on Taiwan was also preparing for military actions to 'counterattack the mainland.' Thus, during the First FYP, these considerations prompted the state to locate 472 of the 694 most important industrial projects in the interior, mainly along the Beijing–Guangzhou

Table 3.10 Investment in civilian key-point development projects of the First Five-Year Plan (1953–57) (million yuan)

	No. of projects	Total investment	
		Amount	*Percent*
Total	98	14261	100.0
Eastern region	26	4449	31.2
Beijing	2	94	0.7
Hebei	5	206	1.4
Liaoning	20	4148	29.1
Central region	56	7902	55.4
Shanxi	7	572	4.0
Neimenggu	3	1035	7.3
Jilin	8	1348	9.4
Heilongjiang	20	2051	14.4
Anhui	1	15	0.1
Jiangxi	3	162	1.1
Henan	9	1177	8.3
Hubei	2	1458	10.2
Hunan	3	84	0.6
Western region	16	1911	13.4
Sichuan	2	86	0.6
Yunnan	1	45	0.3
Shaanxi	7	434	3.0
Gansu	5	1313	9.2
Xinjiang	1	33	0.2

Source: Ma and Wei 1997. Determinants of state investment in China. *Journal of Economic and Social Geography* 88(3): Table 2. Reproduced with permission of Blackwell Publishers Ltd, Oxford.

railroad and in areas to the west of it (Peng *et al.* 1989, p. 25). And it is also largely due to military considerations that the state never invested heavily in the coastal provinces of Fujian and Zhejiang that are so close to Taiwan.

Whereas national security and spatial equality were important factors that affected capital construction investment, the need for the development of coastal industries for economic efficiency was not ignored. The relative importance of these factors and the relationship between the coastal and inland regions in terms of industrial development is most clearly expressed by Mao in his highly influential speech 'On the Ten Great Relationships' delivered in 1956. One of the 10 great issues that Mao discussed was the 'relationship between industry in the coastal regions and industry in the interior.' As this speech mirrors the dominant policy of industrial planning and distribution at that time, the relevant passages are quoted below:

> In the past our industry was concentrated in the coastal regions. By coastal regions we mean Liaoning, Hopei [Hebei], Peking [Beijing],

> Tientsin [Tianjin], eastern Honan [Henan], Shantung [Shandong], Anhwei [Anhui], Kiangsu [Jiangsu], Shanghai, Chekiang [Zhejiang], Fukien [Fujian], Kwangtung [Guangdong], and Kwangsi [Guangxi]. About 70 percent of all our industry, both light and heavy, is to be found in the coastal regions and only 30 percent in the interior. This irrational situation is a product of history. The coastal industrial base must be put to full use, but to even out the distribution of industry as it develops we must strive to promote industry in the interior... not so long ago, there was still fighting in Korea and the international situation was quite tense; this could not but affect our attitude towards coastal industry. Now, it seems unlikely that there will be a new war of aggression against China or another world war in the near future.... It would therefore be wrong if we still fail to make full use of the plant capacity and technical forces of coastal industry.... Without doubt, the greater part of the new industry should be located in the interior so that industry may gradually become evenly distributed; moreover, this will help our preparations against war...
>
> (Mao 1956, pp. 286–7)

Significantly, while Mao was calling for industrial development in the interior, he did not neglect the importance of the old industrial bases in the coastal region. Such concerns for national defense, for the development of the backward inland areas, and for economic efficiency were also identified as the primary goals of the First FYP (State Council 1956). The results of implementing these goals are seen in the decline of the share of CCISOE in the eastern region: from 51 percent in 1953 to 41.1 percent in 1964, while the combined share of the inland regions rose from 49 to 58.6 percent.

It should also be noted that investment had risen during the Great Leap Forward (GLF). The rise of investment was related to the initiation of the policy of establishing 'local independent industrial systems' in 1958, which resulted in the hasty construction of some 216,000 industries throughout the country during the GLF (Liu 1984, 270–71). Provincial and local industrial systems were expected to be self-sufficient in everything they needed. The national frenzy to produce iron and steel during the GLF prompted many regions to build local steel mills even though iron ore or coal was not readily available locally. Poor planning, hasty design and construction, low technology, long-distance transport of raw materials and repetitive construction were only some of the problems that ultimately led to the failure of the GLF, which was followed by a quick decline of investment.

In the winter of 1960, the economic policy of 'adjustment, consolidation, enrichment, and improvement' was announced to solve the problems that resulted from the GLF (Xue 1982). The government decided to reduce the size of investment substantially and to retrench capital construction by suspending many unfinished projects. Per capita CCISOE in the following three years was at a very low level and regional differences were relatively small.

3.4.2 Regional investment during the Third Front period, 1965–71

The pattern of investment in 1965 shows that preference was shifted to the interior provinces, including Gansu, Qinghai and Ningxia (Fig. 3.5). The traditionally poor southwestern provinces of Guizhou and Yunnan also received relatively high per capita investment, but most of the eastern provinces, especially Zhejiang and Shanghai, obtained low per capita CCISOE. Such a pattern is a consequence of the implementation of the Third Front construction program that shifted investment to the inland provinces.

One of the most dramatic and yet least-known episodes in the history of modern China was the Third Front construction program, during which the state allocated a massive amount of resources to developing defense-oriented industries, many from scratch, and to creating self-sufficient industrial base areas in the remote southwestern and northwestern provinces (Naughton 1988). This large-scale, purposive, centrally directed, but poorly and hastily implemented program with shoddy outcomes, was initiated after Mao and the top leadership declared in May 1964 that China should 'watch the development of international situations, prepare for wars,' and 'strengthen the construction of the Third Front areas' (Li and Li 1986).

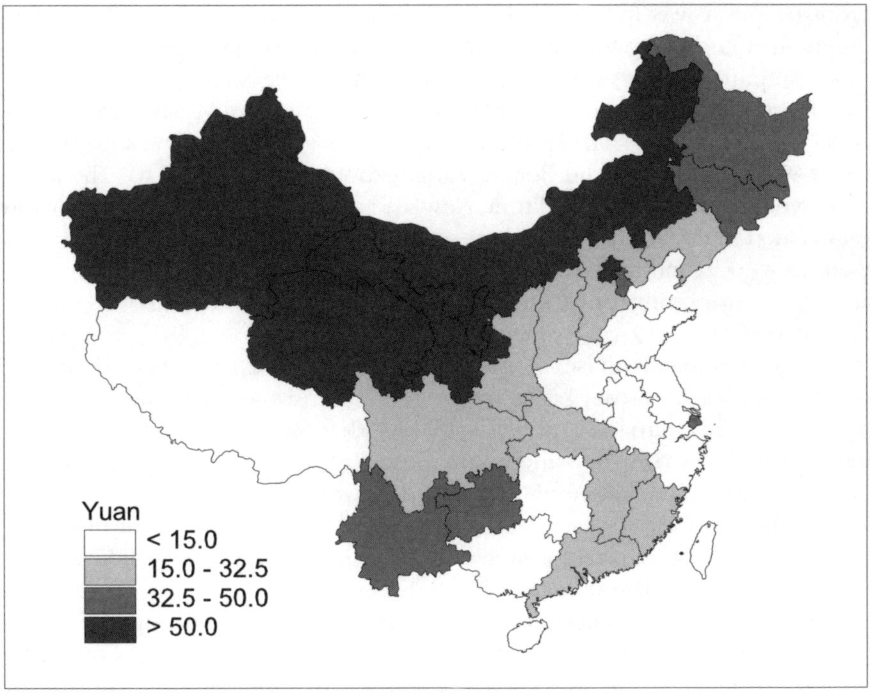

Figure 3.5 Per capita capital construction investment in state-owned enterprises, 1965.

Source: SSB 1990.

China's perception of threats of foreign attacks at that time was not new. It represented a long-term concern for national security that began with the Korean War in the early 1950s during which China's northeastern provinces were potential targets of American bombing. The threats lasted for more than a decade, and at times China responded by threats of her own or by taking military actions, such as during the severe bombing of Taiwan-controlled islands of Jinmen (Quemoy) and Mazu (Ma-tsu) in the late 1950s. As ideological hostility and territorial disputes between China and the Soviet Union intensified in the early 1960s, China's perception of international threats also grew. Additional concerns for national security came from the signing of a mutual defense treaty between the Soviet Union and Mongolia in 1966 and the ensuing deployment of substantial Soviet troops to Mongolia; from China's border clashes with the Russians; and from increasing American military involvement in Vietnam in the late 1960s. The American presence in Vietnam represented a serious threat to the Chinese, who repeatedly expressed their concern and, by early 1966, some 50,000 Chinese troops were sent to Vietnam. The threats from the United States and the Soviet Union were perceived as real and imminent, and nuclear attacks were viewed as likely (Gurtov and Hwang 1980). It was against such a background of perceived international threats to China's national security that the Third Front program was initiated, which was seen by the Chinese leadership as urgent and necessary for survival in the event of foreign aggression.

To prepare for war, China was divided into three strategic fronts.[13] The First Front refers to the coastal and border areas that would be the first to face attacks in a war. The Third Front included the vast interior region south of the Great Wall and west of the Beijing–Guangzhou railway (Fig. 3.6). The areas in between were the Second Front. New and large-scale industrial bases were established in the interior and a large number of industries and research units were moved inland from the coastal region. By 1971, 380 industrial projects and their supporting units and 145,000 workers had been moved inland (Wang 1986, pp. 349–50).

During the initial phase of the program, emphasis was placed on the construction of industrial bases in the remote and mountainous southwest region. Several railroads and self-sufficient defense-oriented industrial systems were hastily designed and constructed. This was followed by the development of the 'three western' (*sanxi*) areas: western parts of Henan, Hubei, and Hunan. Industrial bases were constructed in the Third Front region to produce weaponry, iron and steel, energy, minerals, machine building, electronics and chemical products.

Among the large projects completed during the Third Front period were the railroads linking Chengdu, Kunming, Guiyang, Chongqing, Changsha, and Xiangfan, iron and steel complexes at Panzhihua in Sichuan and at Jiuquan in Gansu, aluminum refining in Guizhou, and airplane production in the Hanzhong region of southern Shaanxi. Located away from major urban centers, these facilities were widely distributed and hidden in mountainous

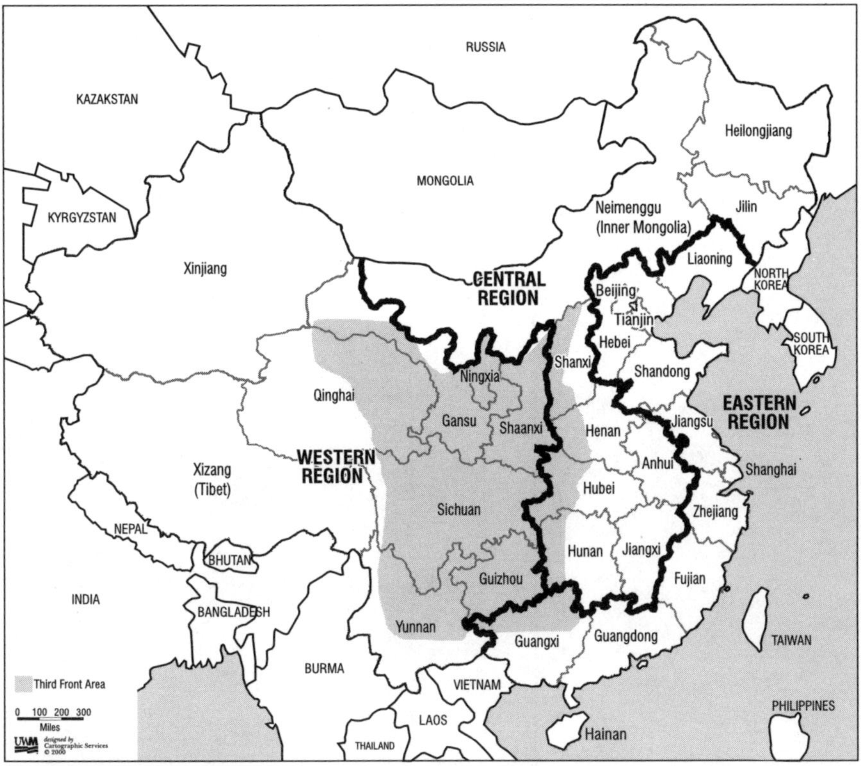

Figure 3.6 Approximate Third Front area.

Source: Based on Ma and Wei 1997. Determinants of state investment in China. *Journal of Economic and Social Geography* 88(3): Fig. 1. Reproduced with permission of Blackwell Publishers Ltd, Oxford.

locations. Their locations were based on the principles of *shan* (in mountains), *san* (dispersal), and *dong* (in tunnels), and they were expected to be dispersed in such a way that they would resemble the gourds of creeping vines (Wang 1990). To minimize possible bombing damage, industrial branches and work shops were located at different sites, and each branch unit was complete with its own transportation systems and living facilities.

Construction of the Third Front projects began in 1965; CCISOE for the western region from 1965 to 1971 was increased dramatically (Table 3.11). Over 2000 large and medium-sized industrial, research and educational projects were quickly built, giving rise to 45 production centers and 30 new industrial cities (*Sanxian diqu*..., 1985). During this period, state investment in these projects totaled nearly 200 billion yuan (Li and Li 1986). In 1965, investment in western provinces began to increase. Investment in the

Table 3.11 Capital construction investment in state-owned enterprises by major periods (million yuan)

	1953–64		*1965–71*		*1972–77*		*1978–90*		*1953–90*	
	Amount	*Percent total*	*Amount*	*Percent total*	*Amount*	*Percent total*	*Amount*	*Percent total*	*Amount*	*Percent total*
Total	185,616	100	137,033	100	196,891	100	1,148,298	100	1,667,838	100
Eastern region	77,269	41.6	41,175	30.1	84,863	43.1	614,606	53.5	817,913	49.0
Beijing	10,599	5.7	4,529	3.3	8,039	4.1	62,698	5.5	85,865	5.1
Tianjin	3,597	1.9	2,002	1.5	5,819	3.0	32,918	2.9	44,336	2.7
Hebei	8,362	4.5	5,101	3.7	11,339	5.8	52,194	4.5	76,996	4.6
Liaoning	17,136	9.2	6,129	4.5	15,840	8.0	73,656	6.4	112,761	6.8
Shanghai	6,588	3.5	3,493	2.5	7,095	3.6	78,574	6.8	95,750	5.7
Jiangsu	6,433	3.5	3,277	2.4	6,515	3.3	58,849	5.1	75,074	4.5
Zhejiang	3,663	2.0	2,024	1.5	3,193	1.6	30,568	2.7	39,448	2.4
Fujian	3,800	2.0	1,723	1.3	2,925	1.5	27,308	2.4	35,756	2.1
Shandong	6,795	3.7	5,014	3.7	9,956	5.1	65,553	5.7	87,318	5.2
Guangdong	7,529	4.1	4,903	3.6	9,459	4.8	109,426	9.5	131,317	7.9
Guangxi	2,767	1.5	2,980	2.2	4,683	2.4	18,349	1.6	28,779	1.7

Central region	66,277	35.7	45,341	33.1	65,202	33.1	327,941	28.6	504,761	30.3
Shanxi	8,014	4.3	5,019	3.7	6,674	3.4	45,967	4.0	65,674	3.9
Neimenggu	6,008	3.2	2,858	2.1	2,913	1.5	27,237	2.4	39,016	2.3
Jilin	6,591	3.6	3,405	2.5	5,539	2.8	26,188	2.3	41,723	2.5
Heilongjiang	12,328	6.6	7,028	5.1	9,671	4.9	57,237	5.0	86,264	5.2
Anhui	5,912	3.2	3,290	2.4	6,039	3.1	30,954	2.7	46,195	2.8
Jiangxi	3,956	2.1	3,402	2.5	3,643	1.9	19,671	1.7	30,672	1.8
Henan	9,586	5.2	5,977	4.4	9,346	4.7	43,162	3.8	68,071	4.1
Hubei	8,084	4.4	8,700	6.3	14,086	7.2	48,623	4.2	79,493	4.8
Hunan	5,798	3.1	5,662	4.1	7,291	3.7	30,472	2.7	49,223	3.0
Western region	42,070	22.7	50,517	36.8	46,826	23.8	205,751	17.9	345,164	20.7
Sichuan	10,897	5.9	18,937	13.8	13,817	7.0	60,836	5.3	104,487	6.3
Guizhou	3,039	1.6	6,593	4.8	4,529	2.3	15,910	1.4	30,071	1.8
Yunnan	4,992	2.7	5,850	4.3	5,395	2.7	21,923	1.9	38,160	2.3
Xizang	212	0.1	320	0.2	638	0.3	4,492	0.4	5,662	0.3
Shaanxi	6,390	3.4	6,653	4.9	8,963	4.6	32,407	2.8	54,413	3.3
Gansu	7,879	4.2	6,317	4.6	6,063	3.1	21,707	1.9	41,966	2.5
Qinghai	2,620	1.4	1,734	1.3	2,282	1.2	14,128	1.2	20,764	1.2
Ningxia	933	0.5	1,457	1.1	1,522	0.8	8,003	0.7	11,915	0.7
Xingjiang	5,108	2.8	2,656	1.9	3,617	1.8	29,288	2.6	40,669	2.4

Source: SSB 1990, 1991, 1997c. Ma and Wei 1997. Determinants of state investment in China. *Journal of Economic and Social Geography* 88(3): Table 3. Reproduced with permission of Blackwell Publishers Ltd, Oxford.

northwestern provinces of Gansu and Qinghai and the traditionally poor southwestern provinces of Guizhou and Yunnan also increased. But most of the eastern provinces, especially Shanghai and Zhejiang, obtained low CCISOE. As Fig. 3.3 shows, the long-term regional patterns of CCISOE during the Third Front period were dramatically reversed, with the eastern region receiving the smallest shares. Such a pattern is a consequence of the implementation of the Third Front construction program, which shifted investment to the inland provinces.

The Third Front program was most intensively implemented during the late 1960s, which overlapped with the most violent years of the Cultural Revolution (1966–69). The late 1960s and early 1970s was also a period in which the policy to develop local industries to help agriculture was implemented. This policy was centered on the 'five small industries' (iron and steel, coal mining, hydropower, machine-building and chemical fertilizers), industries that used local resources and intermediate technology, but the results were far from satisfactory. During this period, political chaos associated with the Cultural Revolution and the overemphasis on defense and local industries seriously undermined the performance of the economy.

The overall shape of CCISOE for the entire study period roughly resembles a dumbbell (Fig. 3.3); its center bar or handle was created during the Third Front. The Third Front program occurred at a time when the violent Cultural Revolution (1966–76) reached its peak during 1966–69. Schools and many factories were closed at the height of the revolution. As noted above, the regional orders of CCISOE were dramatically reversed during the Third Front. The provinces with the highest shares of CCISOE during 1965–71 include Sichuan, Hubei, Heilongjiang, Shaanxi and Guizhou, all in interior China (Table 3.11). But with the abandonment of the Third Front and the tapering off of the Cultural Revolution, the curves of CCISOE for the period of economic adjustment quickly returned to their earlier orders.

3.4.3 Investment during economic adjustment, 1972–77

In the early 1970s, the Third Front program suddenly lost its heat, as reflected by the dramatic changes in the regional patterns of CCISOE shown in Fig. 3.3. Several major events in the early 1970s were responsible for the reversal. Domestically, the fall of Lin Biao, China's most influential leader after Mao and the strongest supporter of the Third Front program, after an alleged coup failed in 1971, seriously undermined the program. This event also coincided with the beginning of China's rapprochement with the United States. It was clear to the Chinese leadership then that the United States had no intention to draw China into the Vietnam War. With her entrance into the United Nations in 1971, China ended its isolation with the West. The signing of the Shanghai Communiqué in 1972 was followed by quick improvements of Sino–American relations. With the changes of leadership in the Soviet Union and the appearance of the 'China card' in the game of international balance of power, the perceived threats of wars vanished and the need for war preparation was quickly reduced. Such developments led to China's decision in the early 1970s to abandon the Third Front program. By the early 1990s, some of the facilities had been moved out of the Third Front areas, while others had closed, suspended operation or shifted their production in favor of civilian products. Some of the units, however, have remained in the Third Front areas. Clearly, fixed investment during the Third Front period was dictated by national security, not by any concern for economic efficiency or spatial equality in development.

The end of domestic chaos and the ease of perceived international threats in the early 1970s prompted the top leadership to change its national policies in favor of open door and economic efficiency. During the period of economic adjustment, the leadership shifted more CCISOE to the eastern region than to the other two regions. In 1972, the state spent 1.1 billion yuan to import 26 large-scale fertilizer and chemical plants from overseas. Fourteen such plants with 54.4 percent of the investment were built in the coastal region, whereas the western region received only eight plants that took only 8.2 percent of the investment (SSB 1991).

Beginning with 1973, no new projects were built in the Third Front region. At the same time, intense efforts were made to develop China's petroleum resources in the coastal provinces, especially the Shengli oil fields in Shandong, the Dagang fields near Tianjin, and the Renqiu fields in central Hebei. In addition, substantial investments were earmarked for the improvement of port facilities along the coast and in the lower reaches of the Yangzi River. From 1973 to 1975, the state invested a total of 2.3 billion yuan to upgrade the ports (Peng *et al.* 1989, p. 212). It is clear from the regional patterns of total CCISOE that the shift to coastal development actually began in the early 1970s, although economic reforms were not formally announced until 1978. By 1978, shares of CCISOE in the western areas were reduced; the coastal provinces of Jiangsu and Hebei were not among the lowest (Fig. 3.7).

3.5 Investment during the reform period

3.5.1 Investment growth and control

High investment rate and endogenous investment cycle are fundamental features of the orthodox command economy (Kornai 1980), including the Chinese economy (Imai 1994).[14] During the reform period, the size of investment has grown rapidly. Fixed investment increased from 96.1 billion yuan in 1981 to 451.7 billion yuan in 1990 and 2297.4 billion yuan in 1995 (Fig. 3.8). While the government has attempted to increase the proportion of income for consumption, the rates of capital accumulation and fixed investment in China remain high. Fixed investment often accounts for 25 percent of GDP, and in 1996, it accounted for 33.5 percent of GDP (SSB 1997c), which is considerably higher than Western countries. Investment decentralization,

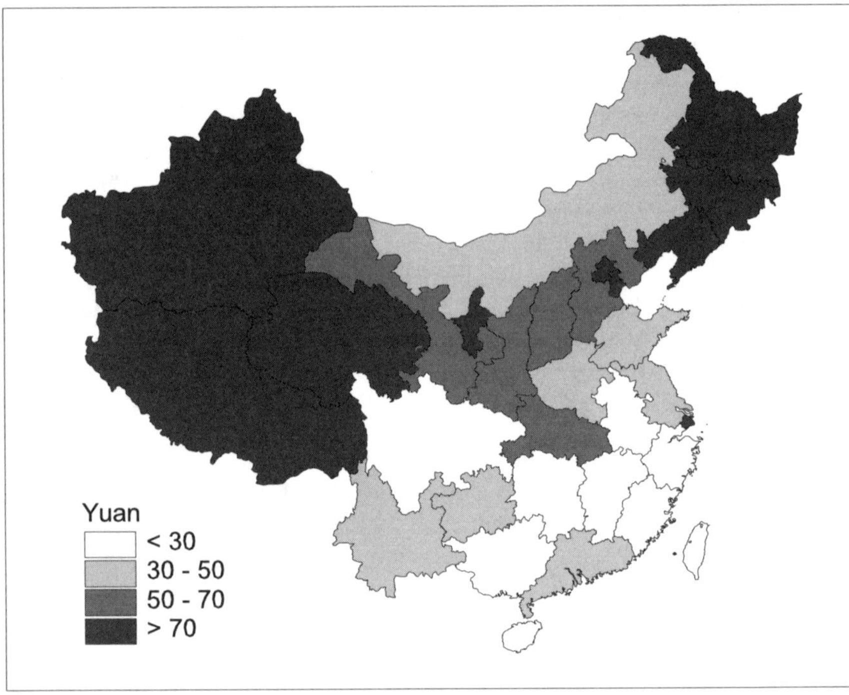

Figure 3.7 Per capita capital construction investment in state-owned enterprises, 1978.
Source: SSB 1990.

rapid growth, high savings rates, and 'investment hunger' have all contributed to the large size and rapid growth of fixed investment in China. Huge investment has been channeled into the Chinese economy, which has contributed greatly to China's phenomenal economic growth; however, it has also placed a huge pressure on inflation, forcing the state to curb investment constantly (Huang 1996).

Local governments, in particular, have actively promoted rapid investment and economic growth, largely for short-term gains and without the consideration of efficiency and sustainable development. Economic growth, which largely determines the advancement of local officials, has been pursued by local governments. The significance of capital contribution to economic growth in China has further prompted the increasing size of investment. Investment funds provided by the banks, largely to the SOEs, were not always subject to rigorous cost–benefit and risk analyses, and this has partially led to huge bad debts owned by state banks (Lardy 1998). Local governments have been heavily involved in the investment process and, partly as a result of this, investment risk and efficiency have not been effectively

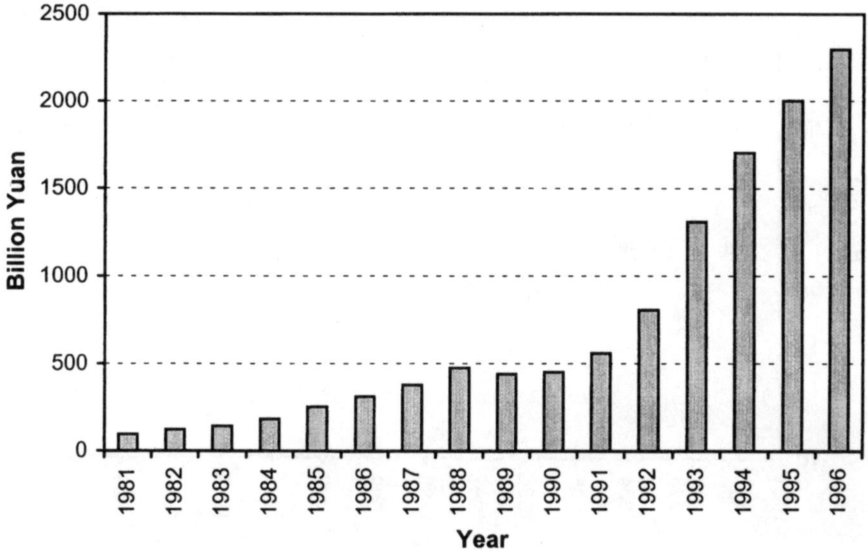

Figure 3.8 The growth of fixed investment, 1981–96 (billion yuan).
Source: SSB 1997c.

taken into consideration in investment decision making (Sun 1988; Jing and Tang 1990).

Rapid investment growth and inflation often forced the central government to control the overheating of the economy by implementing an austerity policy. The austerity policy comprised a mixture of administrative and economic measures (Huang 1996). These included a mandated investment reduction across all provinces, controlling the money supply, bargaining with or even removing local officials, a more rigorous review of the investment process, and cleaning up locally sponsored projects. As shown in Fig. 3.9, growth rates of fixed investment fluctuated greatly over time. The rapid growths of fixed investment in 1984–85, 1988, and 1992–94 were all followed by a slow growth or even decline (e.g., −7.2 percent in 1989) (Fig. 3.9): in 1989, especially, 18,000 projects were closed or postponed (Peng and Huang 1992), which was also partially related to the Tiananmen incident and economic stagnation. To improve investment allocation and management, the central government has introduced a series of measures, including the requirement for project evaluation and feasibility studies, the introduction of the investment orientation tax, and the establishment of state-controlled specialized investment banks. However, the state still largely relies on political and administrative instruments, such as recentralization and credit squeeze, to retain the scale of fixed investment. The central government has been less effective in controlling the volume of investment: during

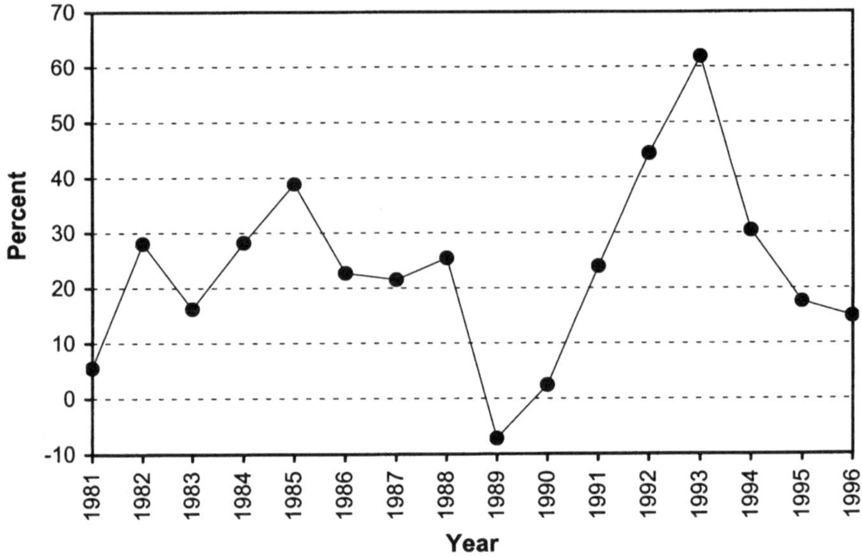

Figure 3.9 Annual growth rate of fixed investment, 1981–96.
Source: SSB 1997c.

the Eighth Five-Year Plan (FYP) period (1991–95), actual fixed investment reached 6.16 trillion yuan, which was 81.3 percent larger than the adjusted Eighth FYP and 106.3 percent larger than the original plan (SPC 1996). Thus, in terms of investment management, China has been facing a dilemma: orthodox approaches of investment control have been less effective and costly, but new approaches appropriate to a transition economy are yet to emerge.

3.5.2 Structural changes of investment

China's reform policies have encouraged the use of multiple sources of investment, including enterprise funds, foreign investment, extrabudgetary funds, and capital from the issuing of bonds and stocks.[15] The structure of fixed investment has changed greatly over time. First, traditional budgetary allocation has been changed to credit allocation in which enterprises rely on bank credits rather than the state budget as a source of capital. During Mao's era, budgetary investment often accounted for more than 80 percent of capital construction investment in the SOEs. During the reform period, with the introduction of market mechanisms and the declining role of budgeting, however, the importance of budgetary investment has decreased dramatically. Budgetary financing of SOE investment declined from 73.2

Table 3.12 Sources of finance for fixed investment (percent), 1981–95

Year	*State budget*	*Domestic credits*	*Foreign investment*	*Enterprise funds*	*Other*
1981	28.1	12.7	3.8		
1985	16.0	20.1	3.6		
1990	8.7	19.6	6.3	52.4	13.1
1993	3.7	23.5	7.3	49.9	15.6
1995	3.1	21.0	11.5	53.2	13.8

Source: SSB 1991–97.

percent in 1978, to 29 percent in 1988 and 5 percent in 1995. From 1981 to 1995, fixed investment financed through budgetary investment declined from 28.1 percent to 3.1 percent (Table 3.12). This change also shows that the state is no longer dominating the investment process, although it still plays an important role.

Secondly, the share of fixed investment in the SOEs has declined, but investment efficiency in SOEs remains poor. Economic returns on fixed assets in the SOEs are lower than any other forms of ownership, especially collective-owned enterprises. Regions, with increasing decision-making power, prefer to allocate their investment in more profitable non-state enterprises. The SOEs are struggling to survive, while township and village enterprises in particular have experienced more rapid investment growth. In addition, the state also intends to adjust the Soviet-type industrial structure by improving investment in energy, transportation, light industry, technology, and services. Even in the SOEs, the share of fixed investment in services increased from 40.7 percent in 1985 to 56.3 percent in 1995. Meanwhile, less than 1 percent of investment went to the primary sector (agriculture).

Thirdly, investment financing beyond state budget and state-controlled banks has increased rapidly, including using enterprise funds and foreign capital. In 1995, state budget and domestic banks only provided 23.6 percent of fixed investment (Table 3.12). As economic reforms allow enterprises to keep some of their profits and fixed capital depreciation funds, more profitable enterprises (often non-state enterprises), most of which are located in more developed coastal provinces, are able to accumulate more capital for investment. This has directly contributed to the rapid increase of investment from enterprise funds, which financed 53.2 percent of fixed investment in 1995. The financing of fixed investment through foreign investment has also increased, and in 1995, foreign capital provided 11.2 percent of fixed investment. As I will elaborate in the next section, enterprise funds and foreign capital are particularly important for the coastal provinces.

In summary, reforms have contributed to investment decentralization, investment growth, and structural changes of investment; problems associated with Mao's investment system have been solved somewhat. New investment resources have been channeled into transportation, technology improvement, services, and non-state sectors. These changes in investment scale and structure are results of central government policies and local initiatives and factors, and the significance of foreign investment in capital formation has also been rising. However, problems still exist in China's investment systems: the nation is still troubled by investment cycles, and investment efficiency remains low, especially in the SOEs.

3.5.3 Investment in the eastern, central, and western regions

Fixed investment in China is uneven among the eastern, central, and western regions and among the provinces. While during Mao's era, regional investment was heavily influenced by government policy, the control of the state over investment has been declining in the reform period, and the influence of local factors and foreign investment in capital formation has risen. More developed and fast-growing coastal provinces have recorded more rapid growth in fixed investment and are less financed by state budget and state banks, while the slowly growing provinces still rely heavily on the state for financing regional development.

Post-Mao reforms have dramatically reduced state control over investment and have provided more favorable policies to the coastal region. Fiscal and investment reforms have allowed regions with more efficient production to channel increasing revenues, profits, and depreciation funds for investment. The coastal region has long established more efficient industrial bases and has advantages in factor endowments, and is therefore quickly capitalizing on new reform policies. The establishment of non-bank financial institutions, such as trust and investment companies, largely in the coastal region, has provided extra sources of funds for fixed investment. The coastal region has also actively bargained with the central government for more preferential policies, such as the coastal development strategy, which favored the coastal region in investment. In addition, coastal provinces have aggressively expanded investment beyond the state budget, especially in extrabudgetary funds, enterprise funds, private capital, and foreign capital. The interior provinces, however, were less favored during the 1980s and early 1990s in terms of investment. They have been less capable of accumulating non-state funds for investment, partly due to poor investment conditions, inefficient production systems, and ineffective local policies.

The concentration of foreign direct investment (FDI) in the coastal region further increases the gap of investment between the coastal and interior regions. Economic reforms have restructured Mao's policy of self-reliance and have promoted foreign investment, with favorable policies in tax, employment, foreign exchange rate, and enterprise management. Special economic zones and open cities have enjoyed the most favorable policy incentives. These policies, as well as local initiatives and networks, facilitated by the globalization of economic activities, have stimulated the rapid growth

Table 3.13 Regional distribution of fixed investment (billion yuan), 1982–95

	1982		*1990*		*1995*		*1982–95 change of share (percent)*
	Volume	*Percent*	*Volume*	*Percent*	*Volume*	*Percent*	
Total	114.5	100.0	433.3	100.0	1937.5	100.0	0.0
Eastern region	62.3	54.4	256.4	59.2	1277.3	65.9	11.5
Central region	34.2	29.9	111.4	25.7	421.0	21.7	−8.1
Western region	18.0	15.7	65.5	15.1	239.2	12.3	−3.4

Source: SSB 1997c.

of FDI in coastal China. The gap between the coastal and interior regions in attracting FDI is astonishing: from the late 1970s to the mid-1990s, China's coastal region attracted more than 85 percent of FDI, while only 1 percent of FDI was allocated to the western region.

As a result of regional differentials in central government policy, local conditions and policies and locational preference of foreign investment, investment disparity between the coastal and interior regions has been growing since the 1970s. From 1978 to 1995, the eastern region's share of total CCISOE increased from 44 percent to 59.1 percent (Fig. 3.3). From 1982 to 1995, the coastal region's share of fixed investment increased from 54.5 percent to 65.9 percent (Table 3.13). In 1995, the coastal region took in 65.9 percent of China's fixed investment, while that in the central and western regions only accounted for 21.7 percent and 12.3 percent, respectively. Per capita fixed investment in the eastern region reached 2790 yuan, much higher than that in the central and western regions (1118 yuan and 1751 yuan, respectively).

The allocation of key-point project investment also represents policy changes of the central government. While during the First FYP period, the state allocated most of the 156 projects to the interior, the emphasis of investment shifted to the coast during the reform period, especially during the 1980s. From 1982 to 1989, the state invested in 319 key-point projects with a total cumulative investment of 310.6 billion yuan; the eastern region received 48.4 percent of this total, whereas the central and western regions were allocated, respectively, 28.4 percent and 12.2 percent (Ma and Wei 1997). During the Seventh FYP period (1986–90), 50.6 percent of key-point project investment went to the eastern region (SSB 1991b). Since the mid-1990s and with the concerns of scholars over rising coastal–interior inequality, the central government has attempted to increase investment in the interior region. However, many interior provinces are troubled by the failure to match the funds provided by the central government (SPC 1997). In addition, as the emphasis of interior investment is on energy, raw materials and infrastructure, investment returns tend to be less efficient and tend to take a longer time to materialize.

3.5.4 Fixed investment in the provinces

While investment has increased in all of the provinces, the provinces with fastest growing shares include Guangdong (4.9 percent), Zhejiang (4 percent) and Jiangsu (2.5 percent), ranked consecutively the first, second, and third, followed by Shanghai (2 percent), Fujian (1.5 percent) and Beijing (1.1 percent), all coastal in location (Fig. 3.10). In 1995, Guangdong alone took in 249.8 billion, which accounted for 12.9 percent of fixed investment, more than the total fixed investment in western provinces (12.3 percent) (Table 3.13). On the other hand, with the declining budgetary investment, many interior provinces, especially China's traditional industrial bases, such as Heilongjiang, Liaoning, Shanxi, and Hebei, experienced slow investment growth and declining shares of fixed investment. The provinces with highest per capita fixed investment are all located in the coastal region, while the provinces of Anhui, Jiangxi, Guizhou, Sichuan, and Gansu had the lowest values (Fig. 3.11).

The coastal provinces are also better positioned in terms of sources of investment. Table 3.14 shows that enterprise funds and foreign investment were more important for the coastal provinces, while interior provinces had higher proportions of investment from the state budget and state-owned

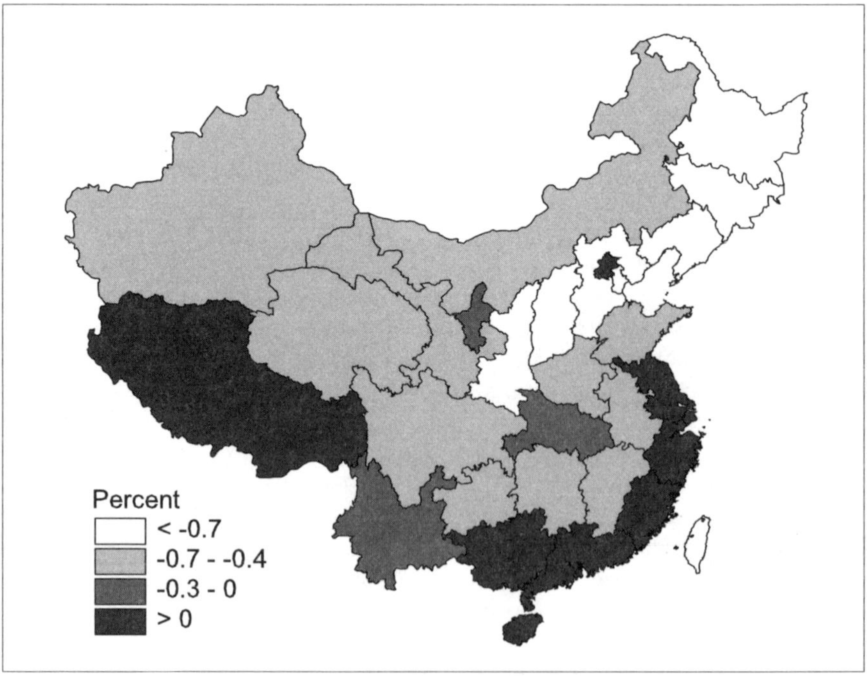

Figure 3.10 Changing shares of the provinces in fixed investment, 1981–96.
Source: SSB 1997c.

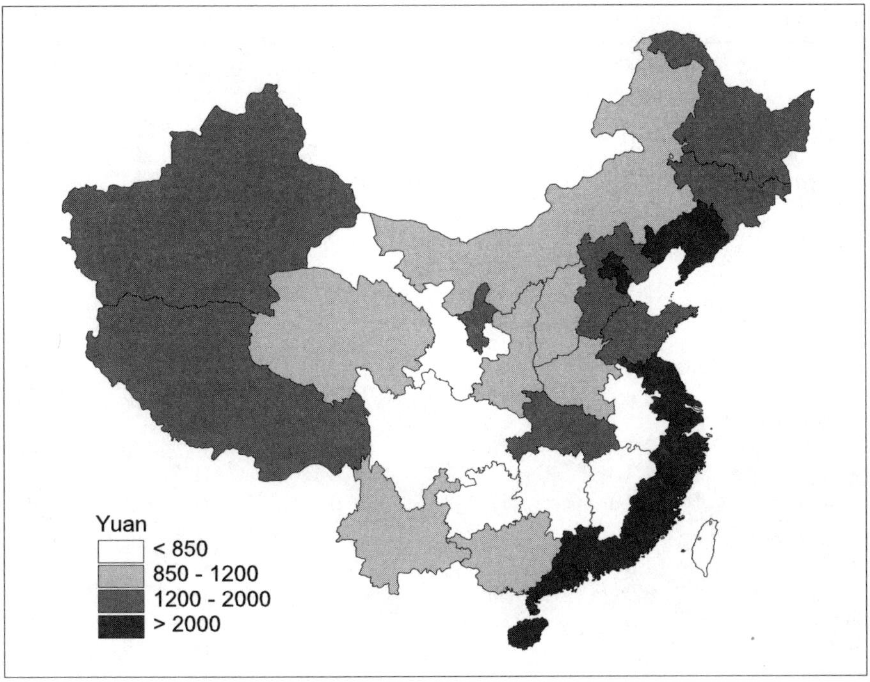

Figure 3.11 Per capita fixed investment, 1995.
Source: SSB 1996.

banks. Budgetary investment and bank credits accounted for more than 50 percent of the CCISOE in interior provinces, while in some coastal provinces, such as Guangdong and Zhejiang, these types of investment only accounted for 15 percent of CCISOE. As shown in Table 3.14, enterprise funds are the single most important source of financing fixed investment. In Zhejiang and Jiangsu, nearly 60 percent of CCISOE is provided by the enterprises for capital construction. In addition, foreign capital provided more funds for CCISOE in coastal provinces, such as 22.8 percent and 33 percent in Guangdong and Hainan, respectively (Table 3.14). However, in several interior provinces, such as Tibet, Qinghai, Gansu, and Guizhou, less than 5 percent of CCISOE was from foreign investment. In other words, interior provinces rely more on the state for funding, which has been shrinking over time and is more rigorously regulated, while coastal provinces finance their economic development mainly through growing enterprise funds and foreign investment.

Moreover, fixed investment in the interior provinces tends to be allocated to less profitable SOEs, while coastal provinces have invested more heavily in non-state enterprises, including TVEs and FIEs. The SOEs in the interior

Table 3.14 Sources of capital construction investment in state-owned enterprises (percent), 1995

	Budgetary investment	*Domestic loans*	*Foreign investment*	*Enterprise funds*	*Others*
China	6.6	22.2	14.3	42.2	12.2
Eastern region	4.8	16.6	16.6	47.9	11.4
Beijing	24.0	14.5	11.7	49.2	7.8
Tianjin	2.9	22.6	25.6	33.8	4.7
Hebei	4.8	14.4	13.5	51.0	15.0
Liaoning	5.4	21.7	17.9	37.0	13.5
Shanghai	2.6	12.4	12.0	53.6	15.1
Jiangsu	5.1	15.3	11.1	58.3	10.3
Zhejiang	2.2	14.4	12.2	60.6	13.2
Fujian	3.4	19.7	14.1	39.0	14.8
Shandong	4.4	23.2	17.0	46.3	8.4
Guangdong	2.0	13.2	22.8	46.0	9.4
Guangxi	4.4	27.7	8.8	41.6	15.2
Hainan	2.8	17.7	33.0	41.4	9.8
Central region	8.0	25.7	14.1	40.6	9.9
Shanxi	4.6	30.8	8.1	51.8	8.4
Neimenggu	8.0	33.5	10.8	35.5	9.1
Jilin	2.2	23.8	27.3	29.0	10.1
Heilongjiang	7.0	19.5	10.5	44.2	14.1
Anhui	8.3	33.6	10.1	37.3	9.8
Jiangxi	10.4	18.0	10.9	42.3	13.3
Henan	8.7	24.5	18.5	38.5	11.3
Hubei	12.5	27.7	15.9	40.2	5.2
Hunan	5.4	20.8	7.6	48.3	11.5
Western region	9.8	29.2	8.0	39.9	10.3
Sichuan	7.7	29.2	9.7	37.6	11.3
Guizhou	6.1	40.3	4.6	40.1	8.7
Yunnan	6.3	20.1	7.5	54.2	8.8
Xizang	68.6	4.0	0.2	19.5	6.7
Shaanxi	15.6	27.9	2.1	42.4	11.2
Gansu	9.9	42.9	7.1	29.0	8.7
Qinghai	9.3	47.1	1.0	26.4	10.6
Ningxia	8.5	32.1	5.7	31.9	14.5
Xinjiang	4.2	29.6	13.3	42.1	10.1

Source: SSB 1996.

have poorer investment returns and have had tremendous difficulty in adjusting to the emerging market mechanisms. Coastal provinces, with less SOEs, have been able to produce more efficient investment returns than the interior provinces. Interior provinces suffered the most during the early 1990s when the central government pushed for market transition and tightened monetary policy.

In summary, post-Mao reforms have dramatically reduced the control of the central government over fixed investment. Regions with more efficient production and more rapid economic growth tend to generate more capital for investment. In addition, the coastal provinces have received more preferential policies and are better positioned in the process of market transition. As a result, the investment gap among China's eastern, central and western regions has increased. The growth rates of investment in the coastal provinces of Guangdong, Zhejiang and Jiangsu are the highest in the nation, whereas many interior provinces, especially traditional industrial bases, lag behind in investment growth. Moreover, investment in interior provinces is more likely from the state budget and state-controlled banks, is largely allocated to the SOEs, and tends to be less efficient, while the coastal provinces are less controlled by the central government and have a stronger ability to accumulate funds through enterprises and foreign investors.

3.6 Implications for regional development

China's fiscal system under Mao was centralized and the center attempted to control the provincial budget. The spending of the central government was mainly for sectoral and strategic reasons with less consideration of regional development (Cannon and Jenkins 1990; Ma and Wei 1997). A large amount of revenues was transferred from the rich provinces to the poor provinces by the central government; however, transferred revenues were not used efficiently, and interior investment has produced poor economic returns (Naughton 1988; Wei 1995). Much of the investment in the interior went to less profitable energy, raw material, and transportation projects. This is particularly true for key projects sponsored by the central government. These projects required a large amount of capital investment, but produced much lower investment returns. Moreover, provinces like Shanghai and Beijing received special treatments from the central government although they were 'upward transfer' municipalities. The center provided them with cheap raw and energy materials, favorable prices for industrial products, interest-free loans, and planned commercial networks, which helped to maintain their leading roles in economic development during Mao's era. However, other coastal provinces generally lacked sufficient funds to improve urban services, to update industrial facilities and to stimulate economic growth.

Post-Mao fiscal reforms have facilitated regional economic development in many ways. China's regional conflicts and uneven development are partially related to the different fiscal systems found in different regions. The central government and local authorities conflict over the level of central control and local autonomy in the bargaining over revenue-sharing rates, expenditure responsibilities, and investment allocation. Favorable fiscal treatment of the central government to some provinces, such as Guangdong, received wide resentment from other provinces: for example, Shanghai, which had heavier upward transfer, had constantly bargained with the center for more fiscal autonomy. These problems were further compounded by the lack of adjustment mechanisms and the heavy involvement of the state in economic activities.

Moreover, decentralization and different economic systems stimulated direct regional conflicts in economic development. Considering themselves unfairly treated by the central government, local governments would use their increased administrative powers to protect their resources and markets. Regional conflicts have been intensified with the outbreak of 'resource wars' such as the 'wool wars' and 'silk cocoon wars' in which resource-rich regions used their powers, even the local branches of the People's Liberation Army, to block the outflow of resources. Regional markets were also heavily protected by local governments that would distort price signals and create unfair competition to limit market penetration of industrial products produced by other regions. Local protectionism and 'investment hunger' prompted regions to duplicate investment projects and to compete for more efficient projects with little consideration of economies of scale and agglomeration.

China's differential fiscal and investment structures have had varied impact on the provinces. The growth rates of Shanghai, Beijing and Tianjin dropped to the lowest positions in the nation during the reform period. The dramatic slowdown in growth is partly related to the erosion of large-scale state support of their economy and the tight central control over local budget. They have had the lowest local retention rates of revenue sharing with the central government. Moreover, whenever the central government lacked funds, these provinces were pressured into providing extra funds for the center. The tight central control and 'upward' revenue transfers have significantly contributed to slow economic growth. Shanghai even experienced revenue decline, which caught the attention of the central leaders, as Shanghai is critical to the coffers of the central government. The 1988 fiscal reform gave these municipalities more autonomy than their previous systems. In the early 1990s, preferential policies were granted to Shanghai, particularly the initiation of the Pudong district, in the hope of revitalizing the city's economy and improving its revenue contribution to the center.

The coastal provinces of Guangdong, Fujian, Shandong, Jiangsu and Zhejiang have registered the fastest income growth since economic reforms, and they are ranked first, second, third, fourth, and sixth, respectively. Rural reforms, open door policies and economic decentralization have stimulated the growth of these provinces. Not only have budgetary funds increased dramatically, but also extrabudgetary funds, which are as important to these provinces as the budgetary funds. They have also actively bargained with the central government to initiate bolder fiscal reforms and to obtain more autonomous fiscal policies. They have benefited the most from investment decentralization, as control over investment resources by the central government has steadily declined while the proportion and the absolute amount of revenues controlled by the local governments have increased.

Investment decentralization has enabled local authorities and enterprises to retain more profits and depreciation funds. Rich coastal provinces are able to keep much of the increasing income that enables them to mobilize new resources for investment. They also actively initiate new investment projects, as they are more competitive than most interior industries. Guangdong and Fujian in particular have very much benefited from fiscal reforms by introducing lump-sum income transfers and obtaining more fiscal powers. Jiangsu and Zhejiang have had less fiscal autonomy and less favorable fiscal systems than Guangdong and Fujian. These provinces argue for more fiscal decentralization and oppose fiscal recentralization. However, they can easily become the targets of central control when economic retrenchment and recentralization are initiated.

However, many interior provinces have been less effective in creating new sources of investment. Western provinces in particular still have to rely heavily on investments from the state budget and state-controlled banks, as SOEs, most of which are losing money, have failed to generate funds for investment. SOEs have poorer investment efficiency and are experiencing slower growth. Non-state sectors, well developed in the coastal region, have generated a considerable amount of investment for growth. The dominance of investment in SOEs in the poor provinces further hindered the economic growth of these provinces.

The border and minority provinces not only retain all of their revenues but also receive transfers from the center. However, economic reforms have dramatically reduced interregional transfers as the center shifts emphasis to economic efficiency and encourages some regions to get rich quicker. These provinces have experienced slower economic structural transformation and lag behind in adapting to market economies. The reduction of transfers and the slow economic transition have led to fiscal distress for many inland counties and provinces. These provinces negotiate heavily with the central government in the hope of obtaining more subsidies and favorable fiscal policies. However, central transfer is limited as the center faces budget deficit itself and is reluctant to provide more assistance to these regions. They also have much less extrabudgetary fund which has become one of the major sources of investment for the coastal provinces.

There is a special group consisting of the provinces oriented to heavy state-owned industries such as Liaoning, Jilin, Heilongjiang, and Shanxi. As economic reforms reduced state subsidy for raw materials and charged heavier taxes for SOEs, provinces with heavy industry as their economic base have been put in a disadvantageous position. Although the central government has raised the revenue retention rates of these provinces and increased the prices of heavy industrial products, they have missed the opportunities created by economic reforms. These provinces have experienced slower economic growth and gradually lost the strong economic positions that they enjoyed in Mao's era. But these provinces can benefit from the reform of SOEs and economic restructuring.

The remaining provinces are in the middle position between the developed and backward regions. They have strong desires to compete with and even surpass the rich coastal provinces. However, their policies have not been as decentralized or as open as the coastal provinces. They are strongly against the uneven regional development strategies of the central government and object to their roles as resource and agricultural bases which the central government has assigned to them. They demand equal access to preferential policies provided to major coastal provinces. The central government has granted them many special policies that were once available only to some coastal provinces. These provinces, particularly those close to the coastal region such as Henan and Anhui, are likely to benefit from the attempt of the central government in the mid-1990s to stimulate reforms and economic growth beyond the coastal region.

3.7 Conclusion

Fiscal and investment reforms are important components of China's economic reforms and have contributed to rapid economic growth in post-Mao China. China's investment system has experienced dramatic changes, including the decentralization of investment management, the introduction of credit allocation, the development of new sources of investment, and the establishment of new measures to manage the investment process. As a result of reforms, more profitable enterprises and coastal regions tend to retain more funds for investment. Private and foreign investments have also increased rapidly in the coastal region, while investment from the state budget and state-controlled banks are more important in the interior provinces. However, the effect of fiscal and investment decentralization is constrained by inefficient governments and regional conflicts. The central government has given some fiscal incentives for the provinces, but it still has not been effective in establishing adjustment systems through the development of markets and the reform of government institutions. Central control over economic activities and the distorted market mechanisms are important factors influencing China's fiscal systems and economic development.

In Mao's era, the central government transferred resources to the poor provinces through the centralized fiscal system. However, resource transfer did not significantly contribute to the development of the poor regions, due largely to the lack of revenue incentives and poor investment returns. Post-Mao fiscal reform has reduced the extent of interregional budget transfer. Although all provinces in China have dramatically increased the size of investment, the gap between the coastal and interior regions in investment has been widening, further compounded by the coastal concentration of foreign investment. The most rapid increase in investment was observed in the coastal provinces of Guangdong, Fujian, Zhejiang, Jiangsu, and Shandong. The post-Mao fiscal system has contributed to the restructuring of the coastal region, i.e. the relative decline of the three municipalities in economic status and the rise of the provinces of Guangdong, Jiangsu, Shandong, Zhejiang, and Fujian. This restructuring is key to the increase of coast–interior inequality and the decline of interprovincial inequality in post-Mao China.

I have shown that investment patterns in China's regions change over time, corresponding to changing domestic and international conditions. Regional investment is neither a cumulative causation process nor a convergent process, but is influenced greatly by the central government policy, local states and agents, and global change. During Mao's era, investment flows across regions were dominated by state funds that were greatly influenced by government policy, while local and international conditions cannot be ignored. During the reform period, while previous regional conditions affect the abilities in capital formation and adapting to new opportunities, state policy, regions, and global investors have all influenced patterns of regional investment. An analysis of state policy, local states and development conditions, and global context, as well as previous patterns, is helpful for a better understanding of investment allocation and regional development.

4 Ownership, foreign investment, and beyond

The non-state sector has grown rapidly, and China is experiencing dramatic changes in shares of output in the state and non-state sectors. This process of changing industrial structure by different forms of ownership (industrial ownership structure) is uneven among China's regions. Besides the role of the central state, regional disparity in ownership structure is heavily influenced by local states and development conditions, such as development experiences and local factors of production. While industrial restructuring, especially deindustrialization, has drawn considerable scholarly attention in developed countries (e.g., Bluestone and Harrison 1982), in China the composition and performance of industries under different ownership greatly influence industrial and regional growth. Regional changes of industrial ownership structure are significantly shaping China's regional development and will be examined in this chapter.

In addition, foreign direct investment (FDI) has increased rapidly, and has become another major force driving regional development in China. Implementation of a series of policies to attract foreign investment is one of the most important objectives of China's economic reforms. China's effort has been facilitated by the globalization of economic activities, during which sizable foreign capital has been invested in developing countries. Much of the research on foreign investment in China has dealt with open door policies, and structure and strategies of foreign investment (e.g., Grub and Lin 1991; Khan 1991; Howell 1993; Luo 1998). Less effort has been made to understand regional patterns and the impact of FDI on regional development in China, which are also to be examined in this chapter.

This chapter analyzes regional patterns of ownership restructuring, FDI, and the impact of these two factors on China's regional development. The chapter also provides a more balanced interpretation of regional inequality in China. In analyzing changes in ownership structure, the importance of past development experiences and the efforts of states in generating regional disparities in ownership structure are emphasized. Also, the importance of states and local conditions in the growth and distribution of FDI in China are shown. Finally, fiscal systems, fixed investment, foreign investment and ownership structure are analyzed to explain the changing patterns of regional development in China in both Mao's era and the reform period.

4.1 Industrialization, ownership, and development

China has experienced rapid restructuring of industrial ownership. However, the reforming of property rights and ownership is not simply privatization or giving the 'right' to private holders (Steinfeld 1998). While SOEs have been struggling to survive, the non-state sector is very much 'growing out of plan' (Naughton 1995a). The Chinese economy is mixed, with different forms of ownership, including SOEs, collective-owned enterprises (COEs), individual-owned (private) enterprises (IOEs), and other enterprises (mainly foreign-invested enterprises (FIEs) and shareholding enterprises (SHEs)).[1] Moreover, regions differ greatly in terms of ownership structure, which has significantly contributed to uneven regional development in China (Wei 1999b). Such disparity is mainly determined by regional/local conditions, including past development experiences, efforts of local governments and agents, and locationally different development conditions, although global forces and national institutions also exert some influences.

After the founding of the PRC, China started socialist transformation to change the means of production. But unlike the Soviet Union, capitalist enterprises in China were transformed first into joint state–private enterprises and then (in the mid-1950s) into SOEs, and some small private enterprises were changed into COEs. From 1949 to 1956, as shown in Table 4.1, SOEs coexisted with COEs, IOEs, and joint state–private enterprises (under 'other' category). Mao believed that in a poor and unevenly developed country like China, socialist transition should be a gradual process and a certain level of collective ownership should be allowed to operate. He referred to these notions of development as 'walking on two legs,' advocating the simultaneous development of industry and agriculture, of light industry and heavy industry, of large enterprises and small enterprises, of state enterprises and

Table 4.1 Gross industrial output by ownership, 1952–95

	Volume (Billion Yuan)				*Proportion (percent)*			
Year	*State*	*Collective*	*Individual*	*Other*	*State*	*Collective*	*Individual*	*Other*
1952	14	1	7	12	41.2	2.9	20.6	35.3
1957	38	13	1	19	53.5	19.0	0.9	26.4
1965	126	14	0	0	90.0	10.0	0	0
1970	186	26	0	0	87.7	12.3	0	0
1978	329	95	0	0	77.6	22.4	0	0
1985	630	312	18	12	64.8	32.1	1.9	1.2
1990	1,306	852	129	105	54.6	35.6	5.4	4.4
1995	3,122	3,362	1,182	1,523	34.0	36.6	12.9	16.6

Source: SSB 1996.

local enterprises, and of enterprises with higher-level and with lower-level technology (Riskin 1987; Morrish 1997).[2] With the completion of the socialist transformation in 1956, SOEs and COEs dominated the Chinese economy. By the end of the First FYP (1953-57), SOEs produced 53.7 percent of the gross value of industrial output (GVIO).

With the initial success of the socialist transition, Mao called for faster economic growth and radical social change by launching the Great Leap Forward (1958-60). Completely unrealistic targets were set for industrial production, and state ownership was preferred (Tang and Ma 1985). Mao and the CCP perceived universal public ownership as a basic goal of socialist revolution and began to seek a higher level of public ownership by further socializing COEs. With the development of SOEs and the transformation of private enterprises, the share of industrial output produced by SOEs increased rapidly, as evidenced by the sudden rise of the share of output in SOEs to GVIO, shown in Fig. 4.1. In 1960, SOEs accounted for 90.6 percent of China's industrial output, while the development of COEs was limited, as COE was considered a transitional and inferior form of socialist economy. The development of COEs was therefore restricted. Moreover, the operation of COEs, especially urban COEs, was largely under state control, but workers employed in COEs earned lower wages and had lower social status. During much of the 1960s, private trading was prohibited and private enterprises were almost totally eliminated, under the slogan of 'cutting the tail of capitalism' (*ge zibenzhuyi weiba*).[3] China began to emerge from its isolation from the West in the early 1970s, and government policies experienced some changes. The hostility towards COEs lessened, and during the 1970s, the share of industrial output produced by non-state sectors began to rise (Fig. 4.1).

During Mao's era, when the operation of private enterprises was prohibited and the state dominated industrial production, China experienced two phases of decentralization that enabled the development of non-state sectors. The first was in the late 1950s during the wave of the Great Leap Forward, and the second was in the early 1970s. Especially during the early 1970s, the central government allowed or even encouraged regions to develop small-scale COEs based on local resources and to serve agricultural production, and regions were given some responsibilities for capital construction, budgeting, and enterprise management. Both phases of decentralization, though ending up in socioeconomic chaos and mismanagement, provided some leeway for local governments and entrepreneurs to establish and manage township and village enterprises (TVEs). As I will elaborate in the next section, these enterprises and the management experiences of local officials, especially those in the coastal provinces, provided some basis for the development of the non-state sector during the reform period.

Since the launch of economic reforms, China's industrial growth has been accelerated and industrial ownership has experienced dramatic changes. During the earlier years of reforms, policies such as rural reforms and

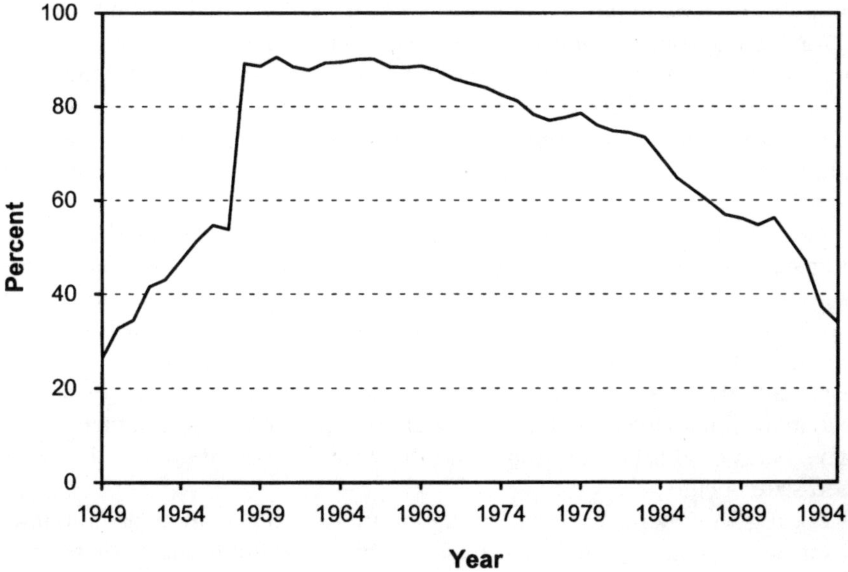

Figure 4.1 Changing shares of industrial output produced by SOEs, 1949–95.
Source: SSB 1996.

open door policies improved the conditions for the development of the non-state sector.[4] TVEs and private enterprises began to mushroom in the early 1980s. Since the mid-1980s, the government has promoted the development of multiple forms of ownership, especially non-state enterprises, motivated by the need to tackle the issue of employment and economic efficiency (Sabin 1994). A series of reforms was introduced in 1984 to promote the development of COEs, and private enterprises were officially permitted to operate. The growth of FIEs was also accelerated. Indeed, reform policies, including financial reforms, investment reforms, marketization and price reforms, reforms of labor and wage systems, and contract responsibility systems, tend to favor the development of COEs. During the 1990s and with the emphasis on market reforms, the growth of non-state enterprises has been phenomenal. Private enterprises have received official recognition and are able to access financial sources such as bank loans. The shareholding system has been encouraged, and the development of SHEs has accelerated. Some TVEs have been restructured by introducing the shareholding system or privatization. All these policies have dramatically changed ownership structure, as the share of SOEs in output and employment has declined further (Fig. 4.1).

SOEs have been more strictly controlled by governments and have to pay higher taxes and administrative fees than enterprises in the non-state sectors. The tax rate for large and medium-sized SOEs was 55 percent. Non-state sectors, on the other hand, have obtained more decentralized and flexible policies and tend to pay lower taxes. FIEs and TVEs benefited the most from favorable tax policies. Only after 1994 were efforts made to equalize sales tax across different forms of ownership at 33 percent. Structural adjustment has been carried out to correct the traditional neglect of consumption and services—a neglect found in most Soviet-type economies—and this has facilitated the development of small-scale services, most of which are in non-state sectors.

The efforts of local states have greatly contributed to the rapid growth of non-state sectors in post-Mao China. As I have discussed previously, non-state sectors, especially COEs, existed during Mao's era. Regions more experienced in developing local economies are keen to establish non-state enterprises. In addition, foreign investors have invested heavily in China, especially in those regions that enjoy favorable open door policies and are rich in local resources. With preferential government policies, numerous FIEs have been established in China, and this has also contributed to the rise of non-state sectors.

Local governments have had an important role to play in ownership transformation. Economic reforms have given local authorities considerable powers in relation to decision making and resource allocation. Economic growth has become the most important objective of local officials, whose personal advancement is based largely on how much economic growth they achieve (Walder 1995b). Reforms, such as decentralization of enterprise management and the introduction of responsibility systems, have given regions considerable power to manage enterprises and to retain increased profits, thus stimulating the efforts of local officials. As a result, local governments are widely involved in industrial production and enterprise management, actively initiate policies that favor local economic growth, and tend to protect the interests of localities (Dwyer 1993). Local governments, particularly township and village governments, are acting as development/entrepreneur states whose motives derive from a desire to maximize income and power. As TVEs and FIEs are less controlled by the central government and are more profitable in operation, considerable resources have been channeled into these enterprises. In short, the efforts of local governments have greatly contributed to the development of non-state sectors in China.

As shown in Fig. 4.1 and Table 4.2, industrial ownership structure in China has experienced profound changes during the reform period. From 1978 to 1995, the share of industrial output produced by SOEs declined from 77.6 percent to 34 percent. Non-state enterprises have grown much faster than SOEs, and in 1995 COEs accounted for the highest percentage of industrial output (36.6 percent). TVEs have been largely under local management and have operated out of state planning. In the 1990s, marketization has provided more decision-making power for local states, enterprises, and individuals. With the efforts of regions and stimulated by foreign investment and trade,

Table 4.2 Industrial output of TVEs and IOEs (billion yuan), 1985–95

		TVEs		*IOEs*	
Year	*Total output*	*Volume*	*Percent*	*Volume*	*Percent*
1985	972	142	14.7	18	1.9
1990	2,392	484	20.2	129	5.4
1993	4,840	1,054	21.8	386	8.0
1995	9,189	2,378	25.9	1,182	12.9

Source: SSB 1996.

Table 4.3 Major efficiency indicators by ownership, 1995

	SOEs	*COEs*	*FIEs*	*SHEs*
Gross output (billion yuan)	2,589	1,584	961	273
Profits and taxes (billion yuan)	287.4	103.9	70.1	36.6
Ratio of gross output to fixed assets (net value)	101.9	288.7	181.6	146.1
Ratio of profits and taxes to fixed assets (net value)	11.3	18.9	13.2	19.6

Source: SSB 1996.

the growth of TVEs has accelerated. From 1985 to 1995, the share of industrial output produced by TVEs increased from 14.7 percent to 25.9 percent, as shown in Table 4.2. Moreover, private enterprises and FIEs also experienced rapid growth, especially during the 1990s. Although the central government prefers state property as the dominant form of ownership, in 1995 IOEs produced 12.9 percent of industrial output. In addition, non-state sectors are less controlled by the central government and are more profitable in operation (Table 4.3). This transformation has fundamentally challenged the traditional notion of socialism and capitalism, as COEs are owned collectively but operated largely outside state planning.

The share of industrial output produced by SOEs has declined dramatically as a result of marketization, globalization and the failure of reforms to revitalize SOEs. By 1995, SOEs only produced 34 percent of GVIO, and the proportions are much smaller in some coastal provinces. Major reform policies have been implemented in an attempt to revitalize SOEs, including changing the system of profit remittance to the tax for profit system (*li gai shui*), administrative decentralization, the establishment of the enterprise contract responsibility system, and the introduction of the Bankruptcy Law (Naughton 1995a).[5] However, as shown in Table 4.3, SOEs are still less profitable. Losses incurred by loss-making SOEs reached 79 billion yuan in 1996. Reforming China's SOEs has not made significant progress due largely to the problem of property right, higher tax and administrative fees, heavy welfare and service charges, price control, the vague role of states in management, and poor risk management and competition abilities (Hodder 1994; Project Group 1994).

4.2 Regional disparities of ownership structure

Patterns of ownership structure differ greatly among China's regions and provinces. In general, the coastal region has lower shares of industrial output in SOEs (Table 4.4). Jiangsu, Zhejiang, Guangdong, Fujian, and Shandong in particular have experienced more rapid growth of non-state sectors and have much lower proportions of output produced by SOEs (Table 4.5). In 1995, SOEs only produced 14.1 percent of industrial output in Zhejiang, and in Jiangsu, COEs produced 60 percent of industrial output.

In order to understand this rapid growth of non-state sectors in the coastal provinces during the reform period it is necessary to look at what happened earlier. First, before 1949, the coastal region produced about 70 percent of China's industrial output, and numerous large and small private enterprises were established there, especially in Jiangsu, Zhejiang, and Guangdong. During the early 1950s, the small enterprises were largely converted into COEs. In 1957, when the transition to socialism had been completed in China, the number of SOEs in Zhejiang province only accounted for 4.6 percent of total industrial enterprises. Secondly, the decentralization in the late 1950s and early 1970s provided some leeway for the development of non-state sectors. These coastal provinces were more experienced with local management and actively captured opportunities to develop local economies. The development of TVEs was especially encouraged by local officials in the countryside, such as rural Jiangsu, as a way of achieving rural industrialization and self-reliance. Thirdly, socialist planning had never totally taken control over the Chinese economy. Even during the Cultural Revolution when the government was hostile to private enterprises, the underground economy could never be totally eliminated. This helps to explain why, in 1970, SOEs in Zhejiang still only accounted for 18.4 percent of industrial units. Another factor to consider is that because many of the large investment projects went to the interior, coastal enterprises were less controlled by the central government and had to improve their abilities to adapt to the constantly changing macro conditions. By 1978, in Jiangsu and Zhejiang, COEs produced almost 40 percent of GVIO.

During the reform period, major policies favored coastal development, including open door policies, fiscal and investment reforms, and administrative decentralization. These uneven development policies to a certain extent have contributed to the more rapid growth of coastal China, especially the growth of the non-state sector (Table 4.6 and Fig. 4.2). With the deepening of reforms in 1992, coastal China has experienced dramatic economic growth and opening up to the outside world. More specifically, several reform policies have contributed greatly to the development of non-state sectors in the coastal provinces.

Table 4.4 Regional shares of industrial output by ownership (percent), 1978–90

	1978		*1985*		*1990*		*1978–90 change*	
	SOEs	*COEs*	*SOEs*	*COEs*	*SOEs*	*COEs*	*SOEs*	*COEs*
China	77.6	22.4	64.9	32.1	54.6	35.6	−23.0	13.2
Eastern region								
Beijing	82.7	17.3	71.1	27.1	63.2	28.9	−19.5	11.6
Tianjin	80.7	19.3	72.1	26.3	59.5	34.6	−21.2	15.3
Hebei	71.9	28.1	58.9	36.8	49.4	39.5	−22.5	11.4
Liaoning	82.4	16.1	67.8	21.7	61.2	28.2	−21.2	12.1
Shanghai	91.1	8.9	77.5	19.8	68.2	19.9	−22.9	11.0
Jiangsu	61.8	37.9	40.5	56.3	34.3	58.0	−27.5	20.1
Zhejiang	60.3	39.7	35.6	62.0	31.2	60.1	−29.1	20.4
Fujian	74.4	25.9	57.0	35.2	45.1	31.4	−29.3	5.5
Shandong	67.7	26.6	58.1	30.2	41.4	50.7	−26.3	24.1
Guangdong	67.9	25.5	53.3	30.0	41.0	34.1	−26.9	8.6
Guangxi	78.1	17.8	75.6	18.3	72.2	20.5	−5.9	2.7
Central region								
Shanxi	73.9	25.2	63.5	31.3	59.7	32.8	−14.2	7.6
Neimenggu	76.5	19.6	78.6	17.3	77.3	18.4	0.8	−1.2
Jilin	78.0	21.2	72.4	25.0	70.4	22.5	−7.6	1.3
Heilongjiang	83.2	15.3	79.5	17.5	80.5	16.5	−2.7	1.2
Anhui	79.4	17.8	64.6	22.4	58.2	31.9	−21.2	14.1
Jiangxi	78.4	21.6	70.2	23.2	65.3	27.2	−13.1	5.6
Henan	74.0	26.0	66.0	30.7	55.2	34.6	−18.8	8.6
Hubei	77.5	22.5	67.5	29.9	62.3	32.6	−15.2	10.1
Hunan	74.0	26.0	67.2	30.7	64.0	28.7	−10.0	2.7
Western region								
Sichuan	80.1	19.9	69.4	28.1	63.7	28.3	−16.4	8.4
Guizhou	81.0	14.3	79.1	12.1	77.3	14.0	−3.7	−0.3
Yunnan	79.7	18.6	78.4	20.0	76.7	20.0	−3.0	1.4
Xizang (Tibet)	73.3	26.7	64.7	11.8	82.7	14.7	9.4	−12.0
Shaanxi	84.2	15.8	75.0	22.9	68.7	23.2	−15.5	7.4
Gansu	90.7	9.2	88.1	11.9	78.1	18.0	−12.6	8.8
Qinghai	81.8	18.2	79.7	18.8	84.1	13.4	2.3	−4.8
Ningxia	85.7	14.3	78.3	17.4	78.6	17.0	−7.1	2.7
Xinjiang	88.9	10.7	83.8	14.7	80.3	16.7	−8.6	6.0

Source: SSB 1990; 1991.

Note: Guangdong includes Hainan.

First, as I have explained in the last section, macroeconomic policies have favored the development of non-state sectors. While SOEs have been in deep trouble, non-state sectors, with less state control, a less heavy tax burden and aggressive marketing, have expanded rapidly. The growth rate of TVEs was accelerated after the shift of economic reforms to urban and industrial reforms in the mid-1980s, which provided a series of preferential

Table 4.5 Declining regional shares of industrial output produced by state-owned enterprises, 1978–95

	1978	*1985*	*1990*	*1995*	*1978–95 change*
China	77.6	64.9	54.6	34.0	−43.6
Coastal region					
Beijing	82.7	71.1	63.2	53.8	−28.9
Tianjin	80.7	72.1	59.5	33.1	−47.6
Hebei	71.9	58.9	49.4	36.2	−35.7
Liaoning	82.4	67.8	61.2	44.0	−38.4
Shanghai	91.1	77.5	68.2	39.5	−51.6
Jiangsu	61.8	40.5	34.3	21.1	−40.7
Zhejiang	60.3	35.6	31.2	14.1	−46.2
Fujian	74.4	57.0	45.1	18.0	−56.4
Shandong	67.7	58.1	41.4	30.8	−36.9
Guangdong	67.9	53.3	41.0	18.4	−49.5
Guangxi	78.1	75.6	72.2	40.2	−37.9
Central region					
Shanxi	73.9	63.5	59.7	45.9	−28.0
Neimenggu	76.5	78.6	77.3	63.4	−13.1
Jilin	78.0	72.4	70.4	62.3	−15.7
Heilongjiang	83.2	79.5	80.5	66.4	−16.8
Anhui	79.4	64.6	58.2	31.7	−47.7
Jiangxi	78.4	70.2	65.3	52.8	−25.6
Henan	74.0	66.0	55.2	34.0	−40.0
Hubei	77.5	67.5	62.3	38.1	−39.4
Hunan	74.0	67.2	64.0	41.7	−32.3
Western region					
Sichuan	80.1	69.4	63.7	40.3	−39.8
Guizhou	81.0	79.1	77.3	67.0	−14.0
Yunnan	79.7	78.4	76.7	69.4	−10.3
Xizang (Tibet)	73.3	64.7	82.7	73.0	−0.3
Shaanxi	84.2	75.0	68.7	60.8	−23.4
Gansu	90.7	88.1	78.1	66.4	−24.3
Qinghai	81.8	79.7	84.1	83.8	2.0
Ningxia	85.7	78.3	78.6	68.7	−17.0
Xinjiang	88.9	83.8	80.3	73.5	−15.4

Source: SSB 1990; 1991; 1996.
Note: Guangdong includes Hainan.

policies for TVEs. The growth rate of private enterprises has increased since the mid-1990s after the government emphasized market reforms and reduced the barriers for the development of private enterprises. In addition, non-state sectors tend to be less regulated and are more flexible in responding to demand. The tax evasion of private enterprises is widespread, although local levies on private enterprises are common (Odgaard 1992). These differences in policy facilitate the growth of the non-state sector. Coastal provinces

Table 4.6 Regional gross value of industrial output by ownership, 1995

	Volume (billion yuan)					*Share (percent)*			
	Total	*SOEs*	*COEs*	*IOEs*	*Other*	*SOEs*	*COEs*	*IOEs*	*Others*
China	9189.4	3122.0	3362.3	1182.1	1523.1	34.0	36.6	12.9	16.6
Eastern region	6065.4	1658.0	2397.6	712.3	1297.5	27.3	39.5	11.7	21.4
Beijing	190.9	102.6	39.5	1.8	46.9	53.8	20.7	1.0	24.6
Tianjin	209.4	69.2	67.3	5.9	67.0	33.1	32.2	2.8	32.0
Hebei	399.6	144.7	150.0	67.3	37.6	36.2	37.5	16.9	9.4
Liaoning	497.5	218.9	150.2	76.3	52.1	44.0	30.2	15.3	10.5
Shanghai	512.9	202.8	95.0	5.7	209.4	39.5	18.5	1.1	40.8
Jiangsu	1181.3	249.8	709.1	52.2	170.2	21.1	60.0	4.4	14.4
Zhejiang	808.8	113.7	373.2	225.6	96.2	14.1	46.1	27.9	11.9
Fujian	280.1	50.3	89.3	43.4	97.0	18.0	31.9	15.5	34.6
Shandong	845.6	260.1	386.6	106.8	92.2	30.8	45.7	12.6	10.9
Guangdong	953.5	171.0	293.6	85.5	403.5	17.9	30.8	9.0	42.3
Guangxi	166.6	67.0	41.9	39.0	18.7	40.2	25.2	23.4	11.2
Hainan	19.3	7.9	1.9	2.9	6.7	40.8	9.6	14.8	34.7
Central region	2188.4	952.2	726.7	355.5	153.9	43.5	33.2	16.2	7.0
Shanxi	175.4	80.5	58.1	32.6	4.2	45.9	33.1	18.6	2.4
Neimenggu	78.2	49.6	14.7	8.5	5.4	63.4	18.8	10.9	6.9
Jilin	142.9	89.1	25.8	12.4	15.6	62.3	18.1	8.7	10.9
Heilongjiang	220.4	146.4	42.0	15.5	16.4	66.4	19.1	7.0	7.4
Anhui	315.6	99.9	136.1	56.2	23.4	31.7	43.1	17.8	7.4
Jiangxi	129.1	68.1	38.1	14.5	8.5	52.8	29.5	11.2	6.6
Henan	471.5	160.1	184.3	94.2	33.0	34.0	39.1	20.0	7.0
Hubei	410.3	156.3	151.8	68.5	33.6	38.1	37.0	16.7	8.2
Hunan	245.1	102.2	75.8	53.3	13.8	41.7	30.9	21.8	5.6
Western region	935.4	511.8	237.9	114.2	71.5	54.7	25.4	12.2	7.6
Sichuan	442.6	178.3	139.1	77.5	47.7	40.3	31.4	17.5	10.8
Guizhou	55.7	37.3	8.4	7.2	2.8	67.0	15.1	12.9	5.0
Yunnan	120.7	83.8	25.2	6.8	4.8	69.4	20.9	5.7	4.0
Xizang (Tibet)	0.9	0.7	0.2	0.1	0.0	73.0	17.8	5.7	0.0
Shaanxi	118.3	71.9	28.2	10.6	7.6	60.8	23.8	8.9	6.4
Gansu	82.5	54.7	15.8	8.3	3.6	66.4	19.2	10.0	4.4
Qinghai	14.9	12.5	1.8	0.4	0.2	83.8	12.0	2.6	1.3
Ningxia	19.8	13.6	2.6	1.1	2.5	68.7	12.9	5.7	12.6
Xinjiang	80.2	59.0	16.7	2.3	2.3	73.5	20.9	2.8	2.9

Source: SSB 1996.

have benefited more from these policies, as non-state enterprises have been more important there than in interior provinces.

Secondly, foreign investment and trade have also facilitated the growth of non-state sectors. China's reforms have encouraged foreign investment and trade through favorable policies in tax, employment, foreign exchange rates, and enterprise management. Special economic zones and open cities, all coastal in location, have enjoyed the most significant policy incentives.

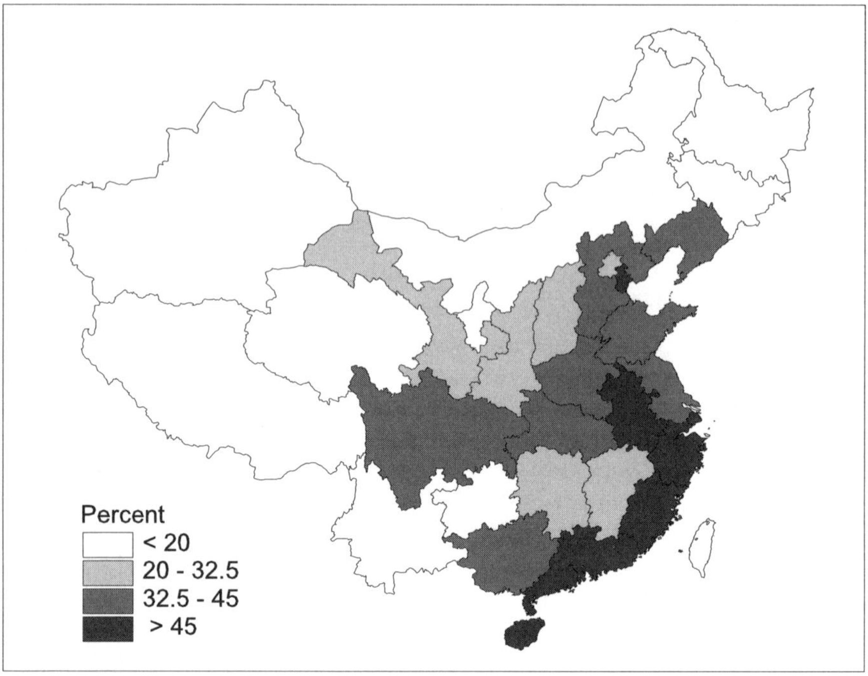

Figure 4.2 Declining regional shares of industrial output produced by SOEs, 1978–95.
Source: Table 4.5.

China's coastal region has generally attracted almost 90 percent of FDI in China. These policies, as well as local initiatives and networks, have stimulated the rapid growth of FIEs, contributing to the growth of the non-state sector.

Lastly, the rapid growth of the service sector in coastal provinces has also contributed to the growth of the non-state sector. China's coastal provinces have experienced more rapid rural industrialization and urbanization, and numerous rural workers have been engaged in services demanded by booming cities and townships. As most services are small scale and SOEs tend to be large scale, non-state enterprises have dominated the service sector. In short, pre-reform experiences, post-Mao reforms, foreign investment and trade, and local efforts have all contributed to the blossoming of non-state enterprises in many of the coastal provinces.

The non-state sector has had higher growth rates and better efficiency than the state sector. In the coastal areas, the non-state sector, especially TVEs, has dominated many rural areas. It contributes significantly to output, employment, and taxes. It provides goods and services that are not covered by the state sector, and also challenges the inefficient and outdated SOEs. The non-state sector has become the most dynamic sector: its phenomenal growth has contributed to the rise of interregional inequality in China.

Those provinces, where a high proportion of industrial output is produced by SOEs, fall into three groups.[6] The first group includes the municipalities of Shanghai and Beijing, which have been more tightly controlled by the central government. Due to the political and economic significance of these provinces, they were given priority in Mao's industrialization policy; a large amount was invested to build them into socialist heavy industrial centers. They also enjoyed preferential treatment, such as cheap energy and raw materials for industrial production. In 1978, SOEs produced as high as 91.1 percent of industrial output in Shanghai. During the reform period, however, SOEs in these provinces have been more tightly controlled by both central and local governments and have failed to benefit from the many opportunities created by the reforms. Their economies experienced slow transformation; and by 1990, SOEs in Shanghai accounted for 68.2 percent of industrial output, much higher than the coastal average (Table 4.5). Only in the 1990s, with the deepening of reforms and the emergence of Jiang Zemin, have more preferential policies been granted to Shanghai, in the hope of revitalizing the economies of China's leading cities. As a result of increased domestic and foreign investments, the growth of non-state sectors in Shanghai and Beijing has accelerated.

The second group includes traditional industrial bases, such as Liaoning, Heilongjiang, Jilin, and Shanxi, which are rich in energy and raw materials. The central government has invested a large amount of resources to build them into heavy industrial bases. During the First FYP period (1953–57), one-third of the 156 key-point projects imported from the Soviet Union were allocated in the northeastern provinces. During the 1960s and 1970s, more investment was channeled into these provinces to strengthen industrial production. SOEs, largely in the heavy industry sector, dominated the economies of these provinces (Table 4.5). For example, in 1978, Liaoning's heavy industry and SOEs produced 73.7 percent and 82.4 percent of industrial output, respectively. Since the late 1970s, while many coastal provinces have promoted the growth of non-state sectors, these provinces have had difficulty in reforming SOEs. In 1990, in Heilongjiang, SOEs produced 80.5 percent of industrial output. Although the share of output produced by SOEs declined rapidly in the 1990s, they still dominated the economies of these provinces (Table 4.5). In 1995, SOEs and heavy industry in Liaoning still produced 44 percent and 75.6 percent of industrial output, respectively, much higher than other coastal provinces.

SOEs in most interior provinces (the third group) were established during Mao's era when there was massive investment in heavy and chemical industry (Auty 1992). During the First FYP period (1953–57), many key-point projects were allocated to these provinces. In the 1960s, a new round of investment was launched to strengthen China's defense industry, most of which was allocated to the interior provinces (the Third Front). Investment under Mao promoted the development of a group of important heavy industrial cities, such as Wuhan, Lanzhou, Baotou, and Panzihua. As a considerable amount of investment went to state-owned heavy and defense industries, SOEs occupied the backbone of the economies of interior provinces. In 1978, SOEs in most of the western provinces produced more than 80 percent of industrial output (Table 4.4). During the reform period, the interior provinces experienced less decentralization than the coastal provinces. Their ownership structures were slow to transform and SOEs still dominated their economies. China's western provinces, in particular, have had difficulty in reforming their SOEs and developing non-state sectors. In 1995, in several provinces, including Qinghai, Xinjiang, Tibet, Yunnan, Guizhou, and Gansu, SOEs still produced more than 65 percent of industrial output (Table 4.6).

Ownership transformation has important implications for China's regional development. The development of non-state sectors is influenced greatly by past development experiences, local state initiatives, and local conditions.[7] During the reform period, non-state enterprises have grown much faster than SOEs and have attracted a considerable amount of investment. TVEs, private enterprises and FIEs have experienced rapid growth, especially during the 1990s. The five coastal provinces of Jiangsu, Zhejiang, Guangdong, Fujian, and Shandong have much lower proportions of employment, output and investment in SOEs. The development of non-state enterprises in these provinces is helping them to grow faster.

SOEs, on the other hand, have lower productivity and have tremendous difficulties in adjusting themselves to the emerging market mechanism. Many SOEs have been operating at a loss because their products are either not needed, of poor quality, or they are not competitive in the domestic and international markets. Many state units operate at reduced capacities, lay off workers, or rely on state handouts for survival. The price reform, marketization, and the drastic reduction of state-subsidized energy and raw materials have greatly increased production costs for SOEs, which play more important roles in the economies of the three municipalities, traditional industrial bases and many interior provinces. In these provinces, higher percentages of investment are absorbed in SOEs that have poorer investment returns than non-state sectors. These provinces have experienced slower economic structural transformation, and the reform of SOEs remains a critical task. While the government has urged the development of TVEs in the interior, the process has been slow and has not achieved much success.

In summary, China has experienced rapid marketization and profound changes during the reform period. The proportion of industrial output produced by less profitable SOEs has declined, while that by TVEs, IOEs, and FIEs has increased rapidly. China's provinces differ considerably in patterns of ownership structure. Some coastal provinces, especially Guangdong, Zhejiang, Jiangsu, Shandong, and Fujian, have experienced more rapid ownership transformation and are in a better position in the development of market economies, while traditional industrial bases and many interior provinces lag behind.

4.3 Growth and structure of foreign investment

Uneven regional distribution of foreign investment is another important factor underlying regional development in China. Foreign investment was very limited before economic reforms, largely due to the Cold War and China's self-reliance policy. During the 1950s, China's major foreign trade and investment partners were Eastern European countries. With the breakup of Sino–Soviet relations in the early 1960s, China, for the following decade, favored self-reliance and was isolated from the outside world. During the early 1970s and with the normalization of US–China relations, China reduced its antagonism to the West, and began to import some advanced industrial equipment, mainly from Japan, the former West Germany, and the United States. Since the late 1970s, China has restructured Mao's policy of self-reliance and favored opening up to the outside world. Attracting foreign investment and promoting international trade are among the major elements of economic reforms. The government believes that such investment can be used to lessen the shortage of capital, advance technology and management skills, promote exports, increase foreign exchange, and stimulate economic growth (Grub and Lin 1991). The importance of foreign investment and foreign-invested enterprises (FIEs) in the Chinese economy has grown rapidly, especially since the mid-1980s when urban reforms and more open door policies were implemented and, more recently, since 1992 when Deng Xiaoping successfully pushed for more economic reforms and an opening door policy.

4.3.1 Policy development and foreign investment

While globalization acts as a broad force driving the rise of FDI at the global scale and in developing countries, investment promotion has come to occupy a prominent place in the development strategies of developing countries and transition economies, and has had strong effects on attracting FDI (Moran 1999). China's policies towards FDI cover a wide range of areas, including:

1. form and control of foreign ownership
2. project duration
3. choice of investment sectors
4. size of investment project
5. choice of investment location
6. tax rates
7. policies on land use
8. management, employment, and wage systems
9. product pricing structures
10. terms for financing and remittance of profits, and other areas (Grub and Lin 1991; Luo 1998).

These policies have had a profound impact on FDI in China. Five stages of the development of state policy and FDI in post-Mao China can be roughly classified.

Initiation (1979–83)

This stage witnessed the gradual opening up of China for foreign investment and trade. To encourage foreign investment, China introduced the Joint Venture Law in 1979, which permitted FDI in China and established principles and procedures for FDI. This was followed by the establishment of four special economic zones (SEZs): Shenzhen, Zhuhai, Shantou, and Xiamen in 1980 (Fig. 4.3), designated by the center and strongly supported by the provinces (Howell 1993). Various state incentives were granted to the SEZs to attract FDI, such as tax and duty exemptions and infrastructure improvement.[8] However, volumes of FDI were small when the approval of FDI was highly centralized, and sectors and regions opened up for foreign investment were very limited (Tables 4.7 and 4.8). During the 1979–83 period, total

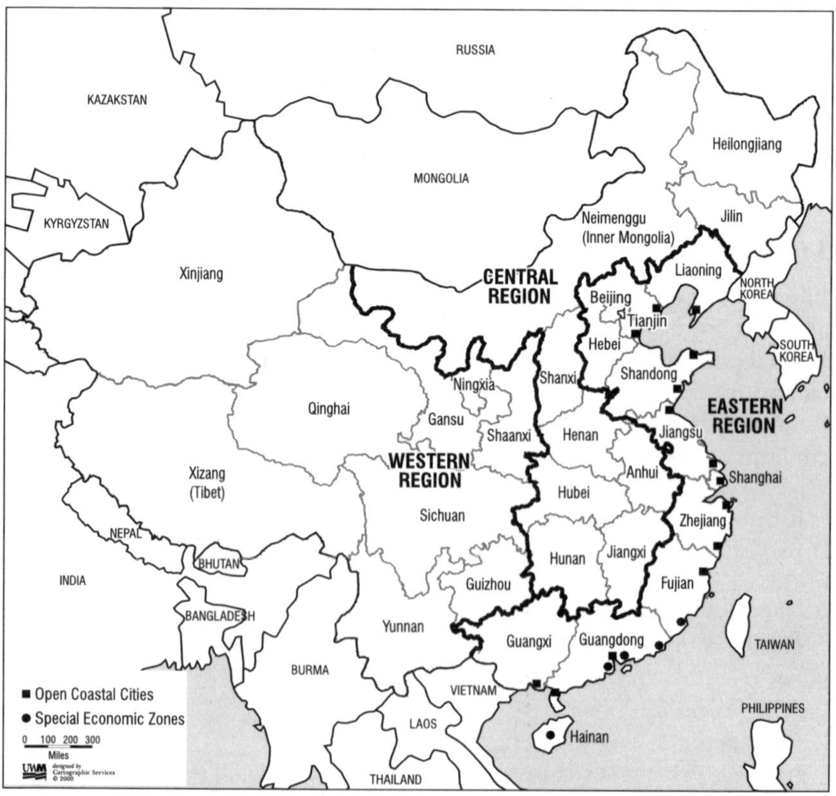

Figure 4.3 Special economic zones, open coastal cities and open areas in China.

Table 4.7 Foreign capital in China, 1979–95 (US$ million)

Year	*Total*	*FDI*	*Foreign loans*	*Other*
1979–83	14,438	1,802	11,755	881
1984	2,705	1,258	1,286	161
1988	10,226	3,194	6,487	545
1990	10,289	3,487	6,534	268
1992	19,202	11,007	7,911	284
1995	48,133	37,521	10,327	285

Source: SSB 1996.

Table 4.8 Foreign trade in China, 1952–95 (US$ billion)

Year	*Total*	*Imports*	*Exports*	*Exports as percent of GDP*
1952	1.9	1.1	0.8	
1962	2.7	1.2	1.5	
1970	4.6	2.3	2.3	
1978	20.6	10.9	9.8	4.7
1985	69.6	42.3	27.4	9.5
1990	115.4	53.4	62.1	16.9
1995	280.9	132.1	148.8	21.4

Source: SSB 1996.

FDI in China was only $1.8 billion, and most investment projects were also small in scale.

Early growth (1984–88)

Corresponding to the shift of reforms from rural reforms to urban reforms and internationalization, China's 'open areas' were extended consecutively from SEZs to 14 open coastal cities (OCCs) in 1984 (Dalian, Qinhuangdao, Tianjin, Yantai, Qingdao, Lianyungang, Nantong, Shanghai, Ningbo, Wenzhou, Fuzhou, Guangzhou, Zhangjiang, and Beihai), and three delta areas (Zhujiang, Minjiang, and Changjiang deltas) in 1985 (Fig. 4.3). OCC policies include three major components (Qu and Green 1997):

1 decentralization of economic decision-making power to OCC governments on the use of FDI to upgrade old or build new production projects;
2 preferential treatments in income tax for technology advanced projects or projects with more than $30 million FDI, and exemption from import and export duties; and
3 the establishment of economic and technological development zones in OCCs.

A series of laws and incentive policies were introduced in the mid-1980s to encourage foreign investment and trade. The Seventh Five-Year Plan (1986–90) announced China's coastal development strategy, which emphasized the internationalization and development of the coastal region as an engine for national development. In 1986, the State Council introduced 'Provisions for Encouraging Foreign Investment' ('the Twenty-two Articles') which offered additional investment incentives.[9] In 1988, two peninsulas (Shangdong and East Liaoning) and a group of coastal cities and counties were opened up for foreign investment, and Hainan Island became a separated province from Guangdong and was designated as China's largest special economic zone. These open areas enjoyed higher foreign exchange retention rates, lower tariffs, favorable taxation for FDI and export, infrastructure improvement, and more decision-making powers. A large number of small and medium-sized foreign trading companies were created, ending the monopoly power of the specialized national foreign trade companies (Lardy 1992). The central government encouraged Taiwanese investors with preferential policies, stimulating FDI from Taiwan in the following years. Moreover, the government approved the operation of wholly foreign-owned enterprises (WFOEs) in some coastal areas. These uneven policies also induced intense competition among regions in an attempt to obtain more preferential policies from the central government.

This period witnessed a rapid opening up of the Chinese economy and an increase of foreign investment in China, as shown in Tables 4.7 and 4.8. However, the amount of FDI remained small, as China's investment conditions—such as infrastructure, labor skills, legal systems, and access to markets—were subject to further improvement. China's frequent retrenchment policies and inadequate regulation regime often frustrated foreign investors (Potter 1995). Although the Chinese government enacted a series of open door and incentive policies, foreign investors remained cautious and were careful in their selection of locations and domestic partners.

Adjustment (1989–91)

In late 1988, China introduced an 'austerity program' to cool down the overheated economy and to reduce the double-digit inflation, which squeezed domestic credit and limited local capital for joint ventures (Sun 1998). After the 1989 Tiananmen incident, China's economic reforms experienced a short period of retrenchment. Meanwhile, many foreign countries boycotted the crackdown by canceling or withholding investment projects and trade contracts. As a result, FDI staggered in 1989 and 1990, and only showed signs of recovery in 1991. Limited improvement in China's open door policy includes the Unified Income Tax Law that eliminated policies restricting the development of WFOEs. Some coastal areas and interior cities were opened up for foreign investment and trade. In 1990, the government approved the establishment of 13 duty-free zones, including Shanghai's Waigaoqiao. Shanghai's

Pudong district was earmarked for foreign investment and trade in 1990, but actual development only took place after 1992. Meanwhile, the Taiwanese government expanded product categories allowed for investment in China.

Rapid growth (1992–95)

In January 1992, Deng Xiaoping toured southern China and pressed for further economic reforms, followed by the Fourteenth Party Congress of October 1992 calling for deeper economic reforms. In this year, 13 border cities, 15 provincial capitals, and five cities along the Yangtze River were opened up. By the end of 1993, China had established 13 duty-free zones and 32 national-level economic and technological development districts. Preferential polices were granted to Shanghai's Pudong district, which quickly became a new center of investment. Gradually, more interior cities and regions, especially border cities, provincial capitals, and cities/counties along the Yangtze River, were opened up to the outside world. By 1996, about a half of Chinese cities and counties were opened up for foreign investment and trade. During this period, China eased restrictions on foreign investment in the service sector. The domestic market was further opened up for foreign products, and tariffs were dramatically reduced. China also established diplomatic relations with South Korea, paving the way for Korean investment, while relations with investors from Taiwan and Singapore were further improved. Moreover, decentralization has brought considerable decision-making powers for local governments, who actively lure foreign capital by providing more favorable policies. China's continuous economic reforms and favorable investment conditions attracted a large amount of foreign investment. FDI increased dramatically from $4.4 billion in 1991 to $37.7 billion in 1995. In addition, investment from leading multinational enterprises (MNEs) and the size of investment projects also increased.

Steady growth (1996–present)

With the rapid growth of the Chinese economy and the emergence of foreign investment, new problems arose. Regions competed intensely for foreign investment by offering numerous incentive policies, sometimes at the loss of state assets and resources, such as the rapid disappearance of agricultural land. Some investment projects only produced pollution and inferior products. FIEs, with ownership advantages and policy advantages such as tax exemption, seriously challenged the SOEs. These problems, intertwined with inflation, corruption, rising inequality, and regional conflicts, prompted the central government to reduce tilted development policies and to pay more attention to social and environmental issues.

In 1995, Beijing released guiding rules for foreign investment that eliminated some preferential policies for foreign investment and trade. Foreign investment projects were classified into four categories: encouraged, permitted, restricted, and prohibited. In 1996, China decided to gradually eliminate preferential tax policies for foreign investment in order to encourage fair competition between domestic and foreign enterprises. The 1998 catalogue classified investments as encouraged, restricted, and prohibited projects. Preferential polices are mainly granted to strategic areas, such as the transfer of advanced technologies, and 'bottle-neck' and basic industries, such as infrastructure, agriculture, environmental protection, and raw materials processing. Beijing seeks not only to direct investment based on sectoral policy, but also to reduce the local approval of redundant or ill-conceived projects, and to improve foreign investment in the interior region (McDaniels 1998). With the change of domestic policies and the influence of the Asian financial crisis, the growth of FDI in China slowed down.

The above analysis has shown that China's open door policy is uneven among regions. Some provinces, especially those in the coastal region, have received more preferential policies from the central government. As I will show in the next section, these uneven policies have directly contributed to uneven distribution of FDI in China, which also relates to differentials in regional endowments and policy initiatives. Moreover, structural patterns of foreign investment are also important to an understanding of regional unevenness of FDI in China.

4.3.2 Structural patterns of foreign investment

FDI in China mainly takes four forms: equity joint ventures (EJVs), contractual joint ventures (CJVs), joint exploration, and WFOEs. CJVs and joint exploration absorbed much of the FDI in China during the early 1980s. EJVs were less significant in the early 1980s but have become the major form of FDI in China (Table 4.9). The upper share limit of foreign investors in EJVs has also been raised over time. With the relaxation of state control over ownership, more WFOEs have been established, firstly in the coastal

Table 4.9 Forms of foreign direct investment (US$ million), 1979–96

	1983		*1987*		*1993*		*1979–96*	
	Total	*Percent*	*Total*	*Percent*	*Total*	*Percent*	*Total*	*Percent*
Total	636	100.0	2,314	100.0	27,515	100.0	176,595	100.0
Equity joint ventures	74	11.6	1,486	64.2	15,348	55.8	90,726	51.4
Contractual joint ventures	227	35.7	620	26.8	5,238	19.0	38,034	21.5
Wholly foreign-owned enterprises	43	6.8	25	1.1	6,506	23.6	42,551	24.1
Joint exploration	292	45.9	183	7.9	424	1.5	5,281	3.0

Source: SSB 1992; MFTEC 1994 and Wang 1997.

Table 4.10 Major sources of foreign direct investment (US$ million)

	1984			*1990*			*1995*		
Rank	*Country*	*Volume*	*Per-cent*	*Country*	*Volume*	*Per-cent*	*Country*	*Volume*	*Per-cent*
1	Japan	1,072	39.6	Japan	3,021	29.4	Hong Kong	20,185	53.4
2	HK	754	27.9	HK	2,432	23.6	Japan	3,212	8.5
3	US	286	10.6	US	596	5.8	Taiwan	3,165	8.4
4	UK	132	4.9	France	579	5.6	US	3,084	8.2

Source: SSB 1992; 1996.

cities. In 1995, EJVs accounted for 43.5 percent of FDI, followed by WFOEs (36.9 percent) and CJVs (19.5 percent). During the 1979–96 period, EJVs took in 51.4 percent of FDI, followed by WFOEs (24.1 percent) and CJVs (21.5 percent).

Major source countries of FDI include Hong Kong, Taiwan, the United States, Japan, Singapore, South Korea, Canada, Australia, and the UK.[10] FDI from Asia has become the dominant source of FDI (Table 4.10). In 1984, among the top eight sources, only Hong Kong and Japan were located in Asia. In 1995, however, except for the United States, the rest of the top eight sources were all Asian countries, which provided 81.9 percent of FDI in China. Shares of FDI from Hong Kong and Taiwan increased rapidly, while that of Japan declined (only 8.5 percent in 1995), reflecting the recession in Japan and the emerging significance of Asian NICs. During the 1979–96 period, FDI from Hong Kong was $103.5 billion, which accounted for 59.1 percent of FDI in China, followed by Taiwan (8.5 percent), the United States (8 percent), Japan (7.9 percent), and Korea (2.1 percent) (Wang 1997).[11]

These Asian countries can be divided into two groups: those dominated by Chinese (Hong Kong, Taiwan, and Singapore) and those neighboring countries with close cultural ties and networks (Japan, South Korea, and Thailand). China has become the largest destination for Hong Kong's overseas investment, and in 1995, Hong Kong alone accounted for 53.4 percent of FDI in China. This indicates the economic restructuring taking place in Hong Kong from manufacturing industry to services, which shifts the labor-intensive industries to China.

FDI in China reflects the trend that multinationals in Asia have emerged as an important force of MNEs. Their preferences for China are due to cultural affinity, government incentives, large potential markets, and favorable labor market conditions. Locational patterns of FDI in China clearly relate to the source composition of FDI. Hong Kong investors prefer south China, especially Guangdong, due mainly to policy incentives, geographic proximity and cultural linkages. Taiwan, Korea, and Singapore also have close cultural ties with China and are provided with favorable policies. Taiwan investors prefer Jiangsu, Guangdong, Fujian and Shanghai, which are close to Taiwan in

Table 4.11 Foreign direct investment in selected sectors, 1995

Manufacturing sector	*Accumulative investment* (US$ *billion*)	*Number of enterprises*	*Assets of FIEs as percent of total*	*Sales of FIEs as percent of total*
Electronics and communications	30.3	2,230	44.5	60.8
Textiles	25.1	4,218	18.8	17.9
Non-metal mineral products	22.4	2,548	16.4	3.4
Electrical equipment and machinery	18.4	2,230	22.4	24.2
Electricity and water	17.4	229	15.5	10.7
Transportation equipment	16.3	149	20.1	25.2
Chemical materials and products	15.8	2,625	12.6	12.6
Garments and fiber products	15.4	5,965	47.9	50.8

Source: Wang 1997, pp. 384–90.

cultural and geographical distances and enjoy favorable local conditions. Considerable amounts of Japanese and Korean investments have gone to north China, including Beijing, Tianjin, Shangdong, and Liaoning. On the other hand, investment projects from the United States and other Western countries tend to be larger in size and to prefer major economic centers and regions, such as Shanghai, Beijing, Tianjin, and the coastal provinces.

The sectoral structure of FDI in China changes over time and, in general, FDI in China tends to be concentrated in manufacturing industries. During the early years of reforms, FDI concentrated in manufacturing industries, real estate, and services. Real estate (especially hotels) absorbed a considerable amount of FDI in China, partly due to restrictive regulations on other sectors and poor infrastructure. Manufacturing investment has become more important since the late 1980s, as market saturation and China's readjustment policy in the late 1980s affected the development of FDI in construction and hotel industries and, increasingly, manufacturing industries are relocated from Asian NICs to China. Although FDI in services has increased since 1992, when China promoted the development of services, manufacturing industries still dominate FDI. Major manufacturing sectors include textiles, garments, chemicals, machinery, transportation, and electronics. (Table 4.11). Labor-intensive industries, through assembly and subcontracting operations, attracted a considerable amount of FDI. FIEs were particularly significant in sectors such as electronics and communications, garments, transportation equipment and textiles.

4.4 Geographic patterns and effects of FDI

Foreign investment tends to be located in major cities and core regions as these places have advantages in labor market conditions, access to domestic and international markets, preferential policy treatment, and infrastructure

Table 4.12 Regional distribution of foreign direct investment (percent), 1983–95

Region	*1983*	*1985*	*1988*	*1990*	*1993*	*1995*
Eastern	92.9	89.6	88.7	93.9	87.5	87.7
Central	1.1	5.2	5.7	8.8	8.8	9.2
Western	5.9	5.2	5.7	3.7	3.7	3.1

Source: MFTEC 1994; SSB 1996.

(Taylor and Thrift 1982). FDI in developed countries experienced concentration and deconcentration processes (O hUallachain 1985). In developing countries, however, overwhelming evidence has shown that foreign capital is 'sticky' and the extent of diffusion to the periphery is limited. Whereas all of the provinces in China have attracted foreign investment, the coastal region has captured the lion's share of FDI.

4.4.1 Uneven geographic distribution

As Table 4.12 shows, a large amount of FDI has been located in China's coastal region, without any significant catching up by the interior. China's coastal region has generally attracted about 90 percent of regional FDI. During the early 1980s, only SEZs and OCCs were opened for foreign investment and trade. With the expansion of open door policies to the interior, FDI began to rise slightly in the interior. As shown in Table 4.12, from 1983 to 1995, the central region's share of FDI increased from 1.1 percent to 9.2 percent, while the western region's share declined. These two regions as a whole attracted only 12.3 percent of China's FDI in 1995. The coastal region has been dominant in the regional distribution of FDI in China.

FDI is unevenly distributed among China's provinces, with Guangdong attracting the largest amount. In 1983, 69.4 percent of FDI was located in Guangdong alone, followed by Beijing (13.1 percent) (Table 4.13). In 1990, FDI in Guangdong still accounted for 45.0 percent of FDI in China. In 1995, Guangdong remained dominant, although its share of FDI had declined to 27.6 percent. The changing distribution mainly took place from Guangdong to other coastal provinces, including Jiangsu, Fujian, Shanghai, and Shandong.

The share of FDI in the interior increased only slightly, mainly in the provinces near the coastal region, such as Hubei, Hunan, Henan, Anhui, and Jilin. Most other interior provinces have not experienced a rapid increase in FDI; some have even registered declining shares. The gap between the coastal and interior provinces in attracting FDI remains large. In 1995, the provinces with FDI over US$1 billion were all located in the coastal region, including Guangdong, Jiangsu, Shanghai, Fujian, Tianjin, Liaoning, Zhejiang, and Beijing (Fig. 4.4). Larger regional gaps also exist in terms of FIEs. In 1995, registered capital in FIEs in the coastal region was US$333.5 billion, which accounted for 84.7 percent of the total (Table 4.14). FIEs in the coastal region

Table 4.13 Provincial distribution of foreign direct investment (US$ million)

	1983		*1990*		*1995*	
	Volume	*Percent*	*Volume*	*Percent*	*Volume*	*Percent*
Total	577	100.0	3,243	100.0	37,215	100.0
Eastern region	536	92.9	3,045	93.9	32,641	87.7
Beijing	76	13.1	277	8.5	1,080	2.9
Tianjin	5	0.8	83	2.6	1,521	4.1
Hebei	1	0.2	39	1.2	547	1.5
Liaoning	7	1.2	248	7.6	1,425	3.8
Shanghai	11	1.9	177	5.5	2,893	7.8
Jiangsu	9	1.5	141	4.3	5,191	13.9
Zhejiang	3	0.4	48	1.5	1,258	3.4
Fujian	16	2.8	290	8.9	4,044	10.9
Shandong	3	0.5	151	4.7	2,689	7.2
Guangdong	400	69.4	1,460	45.0	10,260	27.6
Guangxi	6	1.1	30	0.9	673	1.8
Hainan	0	0	101	3.1	1,062	2.9
Central region	6	1.1	122	3.8	3,429	9.2
Shanxi	0	0	3	0.1	64	0.2
Neimenggu	3	0.6	11	0.3	58	0.2
Jilin	0	0	17	0.5	408	1.1
Heilongjiang	1	0.2	25	0.8	517	1.4
Anhui	0	0	10	0.3	483	1.3
Jiangxi	0	0	6	0.2	289	0.8
Henan	0	0	10	0.3	479	1.3
Hubei	0	0	29	0.9	625	1.7
Hunan	2	0.4	11	0.3	508	1.4
Western region	34	5.9	75	2.3	1,145	3.1
Sichuan	19	3.3	16	0.5	542	1.5
Guizhou	0	0	5	0.2	57	0.2
Yunnan	0	0	3	0.1	98	0.3
Xizang (Tibet)	0	0	—	—	—	0.0
Shaanxi	3	0.4	42	1.3	324	0.9
Gansu	13	2.2	1	0.0	64	0.2
Qinghai	0	0	—	—	2	0.0
Ningxia	0	0	1	0.0	4	0.0
Xinjiang	0	0	7	0.2	55	0.1

Source: SSB 1992; 1996.

also had exports of US$45.6 billion, which accounted for 97.2 percent of the total export.

FDI in China is also unevenly distributed among cities, especially SEZs, OCCs, and cities in the Zhujiang Delta and Yangtze Delta. In 1984, FDI in SEZs reached US$343 million, which accounted for 38.8 percent of the total, and in addition, OCCs attracted 25.4 percent of FDI (Table 4.15). The dominance of SEZs and OCCs in FDI was not challenged during the 1980s; only in the 1990s, shares of these cities in FDI declined somewhat, partially due to

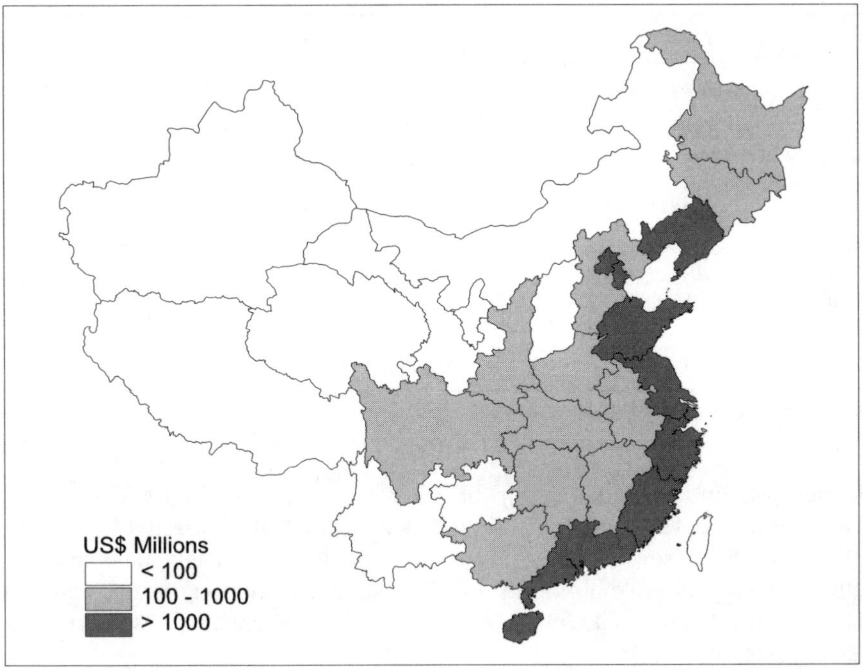

Figure 4.4 Provincial distribution of FDI, 1995.
Source: SSB 1996.

Table 4.14 Foreign-invested enterprises in selected provinces, 1995 (US$ million)

	Number of enterprises	*Total investment*	*Registered capital* *Total*	*Registered capital* *Foreign (percent)*	*Export*
Eastern region	191,555	544,807	333,489	66.1	45,576
Beijing	9,691	28,820	15,115	61.3	709
Shanghai	14,487	67,729	36,029	64.7	4,018
Jiangsu	22,950	53,231	32,404	56.9	2,936
Zhejiang	11,237	24,371	14,806	55.1	1,110
Guangdong	59,582	204,627	129,132	72.0	25,762
Central region	29,184	58,422	41,755	54.2	1,035
Hubei	5,758	13,649	9,278	58.6	187
Western region	11,337	25,955	18,446	55.1	262
Sichuan	5,897	12,680	9,303	53.5	91

Source: SSB 1996.

Table 4.15 Foreign direct investment in SEZs and OCCs (US$ million)

	1984		*1987*		*1990*		*1993*	
	Total	*Percent*	*Total*	*Percent*	*Total*	*Percent*	*Total*	*Percent*
SEZs	343	38.8	343	23.7	646	19.9	3,758	13.7
Shenzhen	184	20.8	274	18.9	349	10.8	989	3.6
Hainan	0	0	0	0	101	3.1	707	2.6
Others	159	18.0	69	4.8	196	6.0	2,062	7.5
OCCs	225	25.4	517	35.7	902	27.8	8,350	30.3
Shanghai	28	3.2	214	14.8	177	5.5	3,160	11.5
Guangzhou	151	17.1	54	3.7	181	5.6	1,278	4.6
Others	46	5.2	249	17.2	544	16.8	3,912	14.2
Regional	885	100.0	1,448	100.0	3,243	100.0	27,515	100.0

Source: SSB 1992; MFTEC 1994.

rising costs and the opening up of other areas for FDI. In 1993, SEZs and OCCs attracted 13.7 percent and 30.3 percent of FDI, respectively. Besides SEZs and OCCs, several other provincial capitals and coastal cities have also attracted considerable amounts of FDI, such as Suzhou, Wuxi, Nanjing, Hangzhou, Fuzhou, Dongguan, Zhongshan, Shenyang, Wuhan, and Chongqing (Fig. 4.5).

Increasing foreign trade is another feature of the post-Mao economy, which is closely related to the development of FDI and FIEs. In 1996, imports and exports of FIEs accounted for 47.3 percent of foreign trade. In terms of regional patterns, the coastal region dominates foreign trade. In 1993, 81.1 percent of China's exports came from the coastal region while those in the central and western regions were only 13.1 percent and 5.8 percent, respectively. Provincial patterns of foreign trade are also spatially uneven (Table 4.16). Guangdong's exports constitute 33.5 percent of China's total, followed by Shanghai (7.1 percent), while in many western provinces, foreign trade was very small.

Several factors have contributed to the leading role of coastal provinces in attracting FDI. First, as I showed in the last section, the central government granted these provinces preferential policies in dealing with foreign investment, currency exchange, and foreign trade. Secondly, these provinces have advantages in location, resources, networks, and local policies. These provinces have locational advantages in foreign investment as well, such as more favorable local policies, more productive economies, a well-developed infrastructure, and a well-trained labor force. They have long-established closer linkages with foreign countries, especially neighboring Asian countries.

Globalization and government policies are important to the growth and distribution of foreign investment. The growth of FDI in China is related to the globalization of economic activities during which FDI has increased at the global scale, including in developing countries. Moreover, the efforts of the

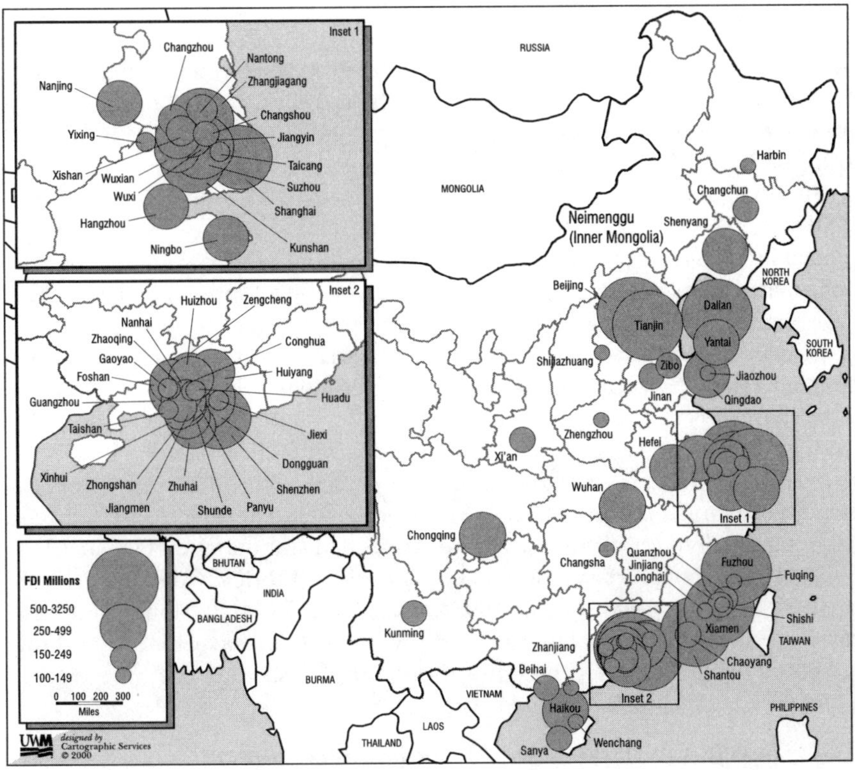

Figure 4.5 FDI in Chinese cities, 1995.
Source: SSB 1996b.

Table 4.16 Foreign trade in selected provinces, 1995 (US$ billion)

	Exports		*Imports*		
	Amount	*Percent*	*Amount*	*Percent*	*Exports as percent of GDP*
China	148.8	100.0	132.1	100.0	21.5
Eastern region	129.7	87.2	115.7	87.6	32.2
Beijing	6.0	4.0	10.8	8.2	35.7
Shanghai	13.1	8.8	12.8	9.7	44.5
Jiangsu	10.1	6.8	8.0	6.0	16.3
Zhejiang	8.3	5.6	4.5	3.4	19.6
Guangdong	59.0	39.7	49.5	37.5	91.6
Central region	13.2	8.9	10.4	7.9	7.0
Heilongjiang	3.0	2.0	1.6	1.2	12.3
Hubei	1.8	1.2	1.9	1.5	6.4
Western region	5.8	3.9	6.0	4.5	6.0
Sichuan	2.0	1.4	1.9	1.4	4.8
Yunnan	1.2	0.8	1.1	0.9	8.1

Source: SSB 1996.

central government have also stimulated the growth of FDI in China. Policies towards coastal provinces, especially SEZs and OCCs, have been the most attractive to foreign investors, which have greatly influenced the uneven distribution of FDI in China.

The concentration of FDI in coastal provinces is also due to the effort of local states in promoting foreign investment. First, local governments in the coastal areas have obtained more decision-making power from decentralization, marketization, globalization, and uneven development policies. They have vigorously bargained with the central government to obtain favorable open door policies. Secondly, they have actively promoted foreign investment and trade by providing numerous preferential policies to foreign investors (Zhang 1994). Local governments act as development states by providing land, infrastructure, and even training labor to attract FDI. Thirdly, local states also act as entrepreneurs who directly negotiate with foreign investors and engage in management and production processes. These acts of local states have significantly contributed to the uneven distribution of FDI in China. Besides local states, coastal provinces also enjoy historically developed social networks (Leung 1996; Hsing 1998). These regions have long-established closer ties with foreign countries and investors; overseas Chinese and their networks have channeled much of the FDI in China.

4.4.2 Foreign investment and regional development

Foreign investment provides capital for economic development, promotes employment, and stimulates urbanization and regional growth. First, foreign investment has become an important source of financing development in China, as evidenced by the increasing proportion of fixed investment provided by FDI: from 3.6 percent in 1985, to 6.25 percent in 1990 and 11.5 percent in 1995. Indeed, contribution of FDI to capital formation is the primary purpose of China's open door policy. Secondly, FDI has contributed to China's technological progress, as FIEs tend to adopt more advanced technology. This form of technology transfer offers institutional and management supports, and is more effective than other forms, such as machinery imports and licensing arrangements (Khan 1991). Moreover, foreign investment has facilitated the process of marketization, structural transformation and the upgrading of domestic industries. Regions with more foreign investment tend to have higher productivity. Thirdly, FDI has stimulated the increase of foreign trade in China and improved China's integration with the global economy. In 1995, FIEs, most of them coastal in location, had a total value of imports and exports of US$109.8 billion, which accounted for 39.1 percent of China's exports. Exports as a percentage of GDP increased from 4.7 percent in 1978 to 21.4 percent in 1995. Lastly, FDI has also contributed to foreign exchange adjustment, the improvement of management, creation of job opportunities, income generation, and government revenues. In 1993, FIEs generated 20.6 billion yuan in profit and tax. With SOEs struggling and problems with private enterprises (e.g., poor working conditions), jobs in FIEs have been pursued by many workers. The dominance of manufacturing industries in FDI also provided numerous job opportunities for rural migrant workers, who otherwise can only work as nannies, street vendors, repairers and construction workers.

FDI has also generated some problems and concerns. As China's open door policies were spatially and sectorally uneven, regions receiving more preferential policies tend to have advantages in attracting FDI and promoting international trade. This unfairness has generated much conflict among regions, among sectors and between central government and local governments. The importance of FDI in revenue and income generation has stimulated the competition among local governments. To attract FDI, local governments have offered more preferential policies by providing more land, improving infrastructure, protecting the interests of foreign investors and forming local–global alliances. Numerous open districts have been established and huge agricultural lands have been converted to urban and industrial use, leading to the rapid decline of cultivable land in China, especially in the coastal provinces where the problem of land shortage is even more serious.[12] In addition, to secure suburban farmland for industrial use, local governments used whatever measures they could to take over the land from farmers, often without proper compensation. Local governments also tend to be more tolerant to labor malpractice and environmental problems: these problems are threatening the agricultural sector and sustainable development in China. The Asian financial crisis has brought renewed concerns on stability when a country is exposed to the global economy and when regulations on the global market are lacking. China's policy adjustment during the late 1990s towards FDI partly reflects the concern of the government over these problems.

As many scholars have shown, the penetration of global forces often leads to the increase of regional inequality in developing countries (Taylor and Thrift 1982; Kasarda and Crenshaw 1991). FDI tends to be urban-biased and located in major Third World cities and core regions, as these places tend to have higher investment returns, are more open to the international markets, enjoy better infrastructure facilities, and have a better-educated labor force. The spatial concentration of FDI often leads to a polarized process of urbanization and reinforces regional inequality. Foreign investment has had a great impact on China's regional development. The Chinese leadership has implemented a series of policies, including the open door policy to promote coastal development. As I have shown, FDI has been concentrated in China's coastal provinces and has become an important source of investment for them.

Because of the geographical concentration of FDI, those with large amounts of FDI tend to benefit greatly. In 1995, foreign capital accounted for a quarter of fixed asset investment in Guangdong and Hainan. During the earlier years, FDI was limited to a few coastal areas such as SEZs, which greatly stimulated the growth of these cities. FDI is still more important in

SEZs such as Shenzhen where it often provides one-third of fixed investment. Meanwhile, many interior provinces have attracted much less FDI. In Qinghai, for example, less than 1 percent of fixed investment was from foreign investment. The coastal region has benefited more than the interior region from the growth of FDI in China. In the case of joint ventures, FDI has stimulated the mobilization of domestic capital for profitable production. The significance of FDI is further compounded by the fact that China's economy is still largely investment-driven. In short, the regional bias of FDI has further intensified uneven regional investment and development.

4.5 Development trajectories of provinces

The development paths of leading provinces are particularly important to the understanding of changing patterns of uneven development in China. As I have analyzed in Chapter 2, China's regions and provinces have experienced different development paths. The coastal region as a whole has experienced more rapid growth than the central and western regions. The coast–interior divide has become more intense in China in the mid-1990s. However, the coastal region is not homogenous. Indeed, regional disparity in the coastal region is much greater than other regions. In 1995, the coefficient of variation (CV) based on per capita GDP was 0.92 in the coastal region, much higher than in the central and western regions (0.3 and 0.26, respectively). While most coastal provinces have experienced dramatic growth (e.g., Guangdong and Jiangsu), some have recorded slow economic growth. In other words, regional development trajectories during the reform period differ dramatically among the provinces in the coastal region. An analysis of the development paths of three provinces, Shanghai, Liaoning, and Jiangsu, can shed more light on the factors underlying uneven development in China.

4.5.1 Growth and restructuring of Shanghai

As I have analyzed in Chapter 2, the change of location quotients shows that during the last four decades, the status of the three municipalities rose under Mao but declined during the reform period, although their income and output levels have always been among the highest in China. It is interesting to note that Shanghai and Beijing were better off during the 1950s. During that decade, as they were China's major industrial and political centers; the central government provided them with preferential policies. Although their status declined in the early 1960s, when Chinese cities experienced severe damage due to the Great Leap Forward, they soon recovered and their relative positions in the Chinese economy were improved.

During the reform period, however, their status declined. From 1978 to 1990, the growth rates of Shanghai, Beijing and Tianjin were among the lowest in the nation. The dramatic slowdown in growth is related to the drastic reduction of state-subsidized energy and raw materials, the tight central control over local budget, the dominance of SOEs (Tables 4.4–4.6) and the slow start of the open door process. The competitive edge of their economies that they had enjoyed for years through large-scale state support of the urban economy had been eroded swiftly by the more rapid growth of several coastal provinces, although since the early 1990s, their positions in the national economy, particularly Shanghai's, has been recovering.

When the PRC was established in 1949, Shanghai was the most developed economic center in China with a strong industrial base and a well-established urban infrastructure. Moreover, as Shanghai was the largest contributor to the state coffers, the state provided the city with generous financial support and low-cost energy and raw materials for its SOEs. In the absence of international competition for its products in the domestic market, Shanghai was able to function as the nation's leading economic center. From 1952 to 1978, the growth rate of per capita national income in Shanghai was 5.9 percent, ranked second in the nation (only behind Beijing). In 1978, Shanghai was the number one contributor to China's national income, industrial output, and revenue.

Economic reforms, however, severely eroded the city's dominant economic role in China's economy. During the 1978–91 period, Shanghai's growth rate of per capita national income was only 5.5 percent, far below the national average (7 percent) and among the lowest in the nation. Its ranking in terms of national income slipped from number one in 1978 to number 10 in 1990 and 1991, although its ranking in GDP and industrial output was higher. Four major factors have contributed to the slower growth of Shanghai during the reform period. First, Shanghai missed the initial opportunity of economic reforms and open door policy (Tian 1996). During the early 1980s, preferential policies were granted to SEZs in Guangdong and Fujian, followed by a series of open door policies in coastal open cities and other regions. These policies have greatly contributed to the dominance of Guangdong in foreign investment and trade. Shanghai was opened up for foreign investment and trade much later than the four SEZs and lagged behind in globalizing its economy. Only in the early 1990s were favorable policies provided to Shanghai.

Secondly, while several coastal provinces enjoyed decentralized fiscal policies, Shanghai has been more tightly controlled by the central government. For example, the 1980 fiscal reform provided provinces such as Guangdong and Fujian with more fiscal powers in revenue sharing and expenditures, but Shanghai, as well as Beijing and Tianjin, shared total revenues with the central government and their revenue sharing rates were adjusted every year. During the 1980s, the amount of Shanghai's financial contribution to the state coffers was not reduced, and Shanghai had the lowest local retention rates of revenue sharing with the central government (White 1989; Jacobs and Hong 1994). In 1990, its expenditure was only 48 percent of its revenue, the lowest ratio in the nation. Per capita revenue even declined from 1541 yuan in 1978 to 1180 yuan in 1990. Shanghai was troubled by tight fiscal control and was in serious fiscal distress.

Thirdly, SOEs dominated Shanghai's economy during much of the reform period, but failed to generate growth in income and jobs. From 1978 to 1995, share of industrial output produced by SOEs in China declined from 77.65 to 34 percent. However, in Shanghai, as shown by Table 4.4, in 1990, 68.2 percent of industrial output was still produced by SOEs. Although policies have been introduced to revitalize SOEs, reforming China's SOEs has not made significant progress. On the other hand, non-state enterprises, such as TVEs in Jiangsu and private enterprises in Zhejiang, have grown much faster than SOEs. The decontrol of price that raised the cost of energy and raw materials, and the reduction of state support for SOEs greatly reduced the city's competitive edge. Such a high share of SOEs placed a heavy burden on Shanghai's economic restructuring.

Lastly, the old and overburdened urban infrastructure and industrial equipment had been deteriorating steadily (Wang 1991). Shanghai's industrial equipment in SOEs was largely outdated. Per capita paved road and living space in Shanghai were lower than the national urban average in 1990 (Yusuf and Wu 1997). Urban policy controlling the growth of large cities limited the expansion of Shanghai. The city has also been slower in adjusting to the emerging market mechanisms, where much of raw materials and products are traded. In short, tight central control, dominance of SOEs, deterioration of urban infrastructure, and the slow start of reform programs have all contributed to the declining status of Shanghai. Patterns of economic growth in Beijing and Tianjin resemble very much that of Shanghai.

The declining status of Shanghai in the national economy eventually caught the attention of the central leaders due to Shanghai's importance to state coffers and economic reform, and its able leaders who quickly rose to the central leadership of China (Cheung 1996). Since the early 1990s and with the emerging power of Jiang Zemin and Zhu Rongji, preferential policies have been granted to Shanghai, in the hope of revitalizing Shanghai's economy and improving its revenue contribution to the center. Tightly controlled financial and investment systems were decentralized. With the opening up of Shanghai's Pudong district, domestic and foreign investments in Shanghai have increased dramatically. Shanghai is recapturing its vigor and is ambitious to become a global city.

During the three years between 1991 and 1993, the city invested 31 billion yuan to improve its urban infrastructure, an amount larger than the total of such investment from 1979 to 1990 (Cai *et al.* 1995). From 1992 to 1995, fixed asset investment increased from 35.7 billion yuan to 160.2 billion yuan, an annual growth of 64.9 percent, much higher than the national average (35.6 percent). In 1996, FDI was 3.9 billion yuan, only behind Guangdong, Jiangsu and Fujian. Shanghai's growth rate of GDP during the 1992–96 period also surpassed the national average. Shanghai's economic growth has been accelerated and its status in the national economy is recovering. However, Shanghai's problems cannot be solved overnight, despite massive domestic and foreign investments. Its development and globalization face intense competition from Hong Kong, Beijing, and other cities in the Yangtze valley.

4.5.2 Development and struggling of Liaoning

Before the establishment of the PRC, the northeastern provinces of China (*dongbei* or Manchuria, including Liaoning, Jilin, and Heilongjiang) were among the provinces with the highest per capita industrial output. Liaoning had already developed a strong industrial base and urban infrastructure. During Mao's era, Liaoning's leading role in China's economy was even strengthened, due largely to Mao's emphasis on industrialization through the development of heavy industry. Liaoning was emphasized due to its rich resources and establishments in heavy industry. A large number of investment projects, particularly in iron and steel industries, were allocated there. During the First FYP period (1953–57), Anshan alone received 52 percent of China's total investment in the steel industry (Xia and Li 1993). Among the 156 key-point projects imported from the former Soviet Union, one-third was allocated in northeastern provinces, and Liaoning alone took in 24 projects.

Liaoning benefited greatly from its resource endowments, industrial bases, and Mao's industrialization policy. Although the significance of Liaoning in the national economy declined during the Third Front period, the economy of Liaoning under Mao experienced a growth rate among the highest in China. From 1952 to 1978, the per capita national income growth rate in Liaoning was 5 percent, ranked fourth in the nation (only behind the three municipalities), and much higher than other provinces of China. Liaoning's economy was dominated by SOEs and heavy industry. By 1978, heavy industry and SOEs produced 73.7 percent and 82.5 percent of industrial output, respectively.

However, the northeastern provinces of China, including Liaoning, have been losing their competitive edges in the post-Mao period. Their growth rates were among the lowest in the nation. From 1978 to 1992, the annual growth rate of per capita national income for Liaoning was 5.7 percent, the fifth lowest in the nation (only higher than Qinghai, Ningxia, Heilongjiang and Tianjin). Liaoning's problems to a certain extent resemble that of Shanghai. First, Liaoning's access to economic reform policies, such as the open door policy, was much more limited than many coastal provinces. The province is characterized by many as having a tendency to '*deng, kao, yao*' (or wait, rely, request) (Freeman 1998). Liaoning's local revenue-retention rate, although higher than the three municipalities, has been lower than other coastal provinces. Although the central government has raised local revenue retention rates for Liaoning and the prices of heavy industrial products, Liaoning and other northeastern provinces have lost the opportunities created by economic reforms and have fallen behind in economic development.

Secondly, due to strict state control and heavy taxes on SOEs, declining subsidies for raw materials, and price control for key industrial products such as steel, provinces with a large amount of SOEs and with heavy industry as their economic bases, such as Liaoning, have been in a disadvantageous position (Ding 1996). In 1995, SOEs and heavy industry in Liaoning produced 41.9 percent and 75.6 percent of industrial output, respectively, much higher than other coastal provinces. These enterprises are tightly controlled by the state and cannot compete with emerging FIEs, TVEs, and even private enterprises. The performance of China's SOEs has been deteriorating and almost a half of SOEs lost money in operation. The losses suffered by SOEs grew from 340,000 yuan in 1977 to 3.79 billion yuan in 1992 (Schueller 1997). Liaoning has been troubled by problems with SOEs and has had difficulty in adjusting to the emerging market economies.

Lastly, Liaoning's outdated industrial equipment and overburdened urban infrastructure are difficult to improve in a short period of time. Most of Liaoning's industrial equipment was built in the 1950s and 1960s. Although investment for equipment and infrastructure updates has increased, much more funds are still needed. Consequently, the northeastern provinces of China experienced slower growth during the reform period and have gradually lost the strong economic positions that they enjoyed in Mao's era.

A comparison of Liaoning with Jiangsu and Guangdong can shed more light on the structural differences between declining and rising provinces (Table 4.17). While the gap in per capita GDP is small, Jiangsu and Guangdong are better positioned in China's transition to market economies. In 1995, SOEs only produced 18.4 percent and 21.1 percent of industrial output in Guangdong and Jiangsu, respectively, while Liaoning's economy was still largely dependent on SOEs that are less profitable and struggle to survive. In addition, Liaoning is lagging behind in attracting foreign investment and promoting international trade. From 1978 to 1995, Liaoning's economy grew much slower than Jiangsu and Guangdong, due partially to its problems with SOEs and its slow pace in marketization and globalization.

The Chinese government during the 1990s has stressed the reform of SOEs in an attempt to rescue the declining state sector, which is considered as the cornerstone of socialism. Numerous reform programs have been announced to revitalize SOEs, including the introduction of the shareholding system.

Table 4.17 A comparison of Liaoning with Guangdong and Jiangsu, 1995

Indicator	*China*	*Liaoning*	*Guangdong*	*Jiangsu*
Population (million)	1,211	40.9	68.7	70.7
GDP (billion yuan)	5,826	279	538	516
GDP per capita (yuan)	4,810	6,829	7,831	7,291
SOE's output (percent)	34.0	44	18.4	21.1
FDI per capita (US$)	31.2	34.8	149.3	73.4

Source: SSB 1996.

Some preferential policies have been provided to improve heavy industry, such as the increase of investment in renovating old industrial equipment. Liaoning has also actively promoted foreign investment and trade, and Dalian even intends to build into a global city. During the 1990–95 period, the annual growth rate of per capita GDP was 9.4 percent in Liaoning, although it is still below national average (14.5 percent). While the restructuring process is slow, Liaoning is adjusting to the emerging market mechanism.

4.5.3 The emergence of coastal provinces

The five coastal provinces (Guangdong, Fujian, Zhejiang, Jiangsu, and Shandong) were not favored by Mao's regime but they are the biggest winners of economic reforms. During much of Mao's era, these provinces were the least favored in development policies. They received few key-point investment projects from the central state due partially to their coastal location, as possible foreign invasion would come from the sea, particularly from Taiwan (Ma and Wei 1997). While Mao did not totally ignore coastal development, the emphasis was placed on traditionally leading industrial bases, particularly Shanghai, Beijing, and Liaoning. During the 1950s with the implementation of the First FYP, which favored the interior and the traditional industrial bases, the status of these provinces in the national economy declined.

Devoid of the preferences of the central government, these provinces developed their strong abilities in using local resources and market forces to improve their economies. The enterprises in these provinces tend to be small and in the non-state sector. In Zhejiang, for example, in 1957 when China completed its socialist transition, the number of SOEs only accounted for 4.6 percent of total industrial enterprises. Their status was improved during the late 1950s with the implementation of the decentralization program and the establishment of some TVEs, and began recovery during the early 1970s when a pragmatic leadership emerged and a decentralization policy was implemented in China for the second time. Non-state enterprises, such as TVEs in Jiangsu and private enterprises in Zhejiang, benefited greatly from decentralization and regional autarky during Mao's era.

Since economic reforms, these coastal provinces have registered the fastest economic growth. Their status has been advanced dramatically, with Zhejiang, Guangdong, and Jiangsu moving up most rapidly, followed by Fujian and Shandong. They have actively bargained with the central government for more decentralized policy and for allowing them to initiate bolder local policies. Guangdong and Fujian in particular have been very much benefited from fiscal reforms by introducing lump-sum income transfers and obtaining higher local retention rates of revenues (Wei 1996). Moreover, SEZs in Guangdong and Fujian were opened up for foreign investment and trade in 1980; this was soon followed by the opening up of OCCs in other coastal provinces. These cities attracted a large amount of foreign investment and international trade surged, particularly in Guangdong (Vogel 1989). Meanwhile, localities have aggressively launched new initiatives and dramatically restructured their economies by developing private and collective enterprises. Services and rural enterprises have flourished there and have become important sources of income growth, especially in Jiangsu and Zhejiang. All these changes have contributed to their phenomenal economic growth in the reform period. Their economic growth was accelerated in the 1990s with the implementation of bolder economic reforms.

The development path of Jiangsu well represents the experiences of these provinces (the following two chapters will analyze the development of Jiangsu in great detail). When the PRC was established in 1949, Jiangsu had already established numerous modern enterprises and was among the most developed provinces of China. During the 1950s, however, the coastal provinces were ignored in industrial allocation, but Jiangsu still had to contribute a huge revenue to the central government. Jiangsu's economic growth in the 1950s was among the lowest in the nation. In the late 1950s, while China experienced chaos and economic crisis, decentralization did give Jiangsu considerable decision-making powers in developing local economies. Some small-scale rural enterprises were established, particularly in Sunan (southern Jiangsu), although most of them were short lived. In the 1960s, heavy industry received a high priority in state investment and the development of light industry was largely overlooked. Light industry stagnated and Jiangsu recorded slow economic growth in the 1960s. During the early 1970s, China's regional development policy began to shift to the coastal region. Meanwhile, during the late 1960s and early 1970s, the central government encouraged the development of rural enterprises. Hundreds of TVEs were established in Jiangsu, especially in Sunan. As a result, Jiangsu's economy experienced a recovery and industrial output surpassed agricultural output. However, during Mao's era as whole, growth rates of major economic indicators were low.

Jiangsu has experienced dramatic economic growth in the post-Mao period. Like other coastal provinces, the province has received favorable development policies from the central government. First, fiscal reforms have provided incentives for the growth of local economies. Jiangsu has traditionally had a heavy revenue burden because of its strong economic base. A new fiscal system of 'fixed rate contracts' was implemented in Jiangsu in 1977 which provided the province more fiscal autonomy and incentives, followed by a series of fiscal systems that provide more fiscal incentives. Under these new systems, Jiangsu could retain an increasing portion of growing revenue. Secondly, economic reforms have given localities greater autonomy in investment decision making, including sources of capital and investment allocation strategies. Regions have greater access to investment from different sources, such as bank credits, private investment, and foreign capital. Local authorities are also motivated to investing in more profitable regions and sectors to achieve higher economic returns. Thirdly, Jiangsu had a larger proportion of non-state enterprises to begin with, and is therefore quickly capitalizing on the reform policies that favor the non-state sector. The development of Jiangsu, particularly Sunan, has also benefited from the markets in and technical know-how provided by nearby cities, particularly Shanghai, Suzhou, Wuxi and Changzhou. Lastly, Jiangsu is among the coastal provinces of China that enjoyed preferential open door policies and increasing foreign investment and international trade. In short, enterprise reforms, fiscal and investment reforms, open door policy and many other reform programs have facilitated the growth of Jiangsu.

4.6 Understanding regional inequality in China

4.6.1 Understanding regional inequality under Mao

Previous analyses have shown that while some efforts were made to develop China's interior provinces, regional inequality based on national income even increased or at least persisted. As temporal trends of CVs and LQs show, during the 1950s, regional inequality did not rise, partially as the result of the implementation of the First FYP. But since the late 1950s, the coastal region has grown more rapidly than the interior region, and interprovincial inequality also rose. In general, regional inequality persisted during Mao's era, especially throughout the 1960s and 1970s; this was caused by several factors. First, China is a nation with diverse development conditions and a historical legacy of regional imbalance. The coastal region has better physical and human conditions for development (Pannell and Ma 1983). Historically, China's regional development was extremely uneven, created partly by the penetration of colonial forces. Before the establishment of the PRC in 1949, industry had been more advanced in the coastal region. As development experiences of many developing countries have shown, this core–periphery structure greatly affects the future development paths, and tends to persist over time.

Secondly, regional equality was not the top priority goal: rather, considerations for industrialization and national defense were even more important than equity in resource allocation. Although efforts were made to balance industrial distribution in the First FYP period, such efforts were de-emphasized after the mid-1950s. Under the influence of the Soviet-model of industrialization, the leading industrial centers, particularly Shanghai, Beijing, and Liaoning, received more favorable policies than other provinces. Coastal development was not neglected by the Maoist regime, but efforts were made to fully utilize the resources of traditional industrial bases (three municipalities and Liaoning in particular). This was considered necessary as these provinces provided huge revenues that were needed for China's industrialization, national defense, and interior development. The government provided them with large state investment, cheap energy and raw materials, and many other preferential policies. As a result, the three municipalities and Liaoning recorded growth rates among the highest in the nation. The rapid growth of these leading provinces contributed significantly to the rise of regional inequality in China under Mao.

Investment under Mao had given rise to a number of industrial cities in the interior. However, interior investment produced poorer economic returns than the coast, particularly during the 1960s. The defense-oriented Third Front project was poorly planned and widely scattered in remote locations with backward infrastructure. They took a long time to build, tied up a great amount of investment, and yielded poor economic results (Naughton 1988; Wei 1995; Ma and Wei 1997). Investment in the interior was also absorbed in traditional industries such as heavy industry, which produced cheaper priced goods and was less efficient in production, compounded by the problem of the low allocative efficiency of the socialist system (Auty 1992). Moreover, a large percentage of interior investment in civil projects went to the improvement of costly transportation and communication facilities. As revealed by Wei (1995), the relationship between state investment and industrial output was extremely poor in the 1960s when large amounts of state investment went to the interior. Poor economic returns of state investment greatly hindered interior development. Investment in the coastal region, however, was allocated to more profitable industries and enjoyed much higher returns, although the development of many coastal provinces, such as Zhejiang and Fujian, was largely ignored by the central state. The low efficiency of interior investment partially explains why the interior fell behind and regional inequality persisted.

Thirdly, Mao's decentralization policy, problems in socialist planning and local forces also contributed to the persistence of the coastal–interior inequality. Decentralization policies, which were implemented during the GLF and early 1970s, provided considerable decision-making powers to the regions. The coastal provinces, such as Jiangsu, were able to capture these opportunities and took advantage of their strong agricultural bases, skillful manpower, ties to the cities, and large markets to develop their economies. Particularly during the early 1970s, with the change of China's macro economic policy, numerous collective enterprises, especially TVEs, were established in the coastal provinces. As a result, the economies of the coastal provinces performed better than most interior provinces.

Unlike the former Soviet Union, the Chinese government had transformed a large amount of private enterprises into collective enterprises, rather than simply SOEs. Even during the periods of tight central control, regions could still make some efforts to protect local economies, as Chinese local economies had never been fully transformed into 'pure' socialist planned economies. However, development policy in the poorer provinces was less decentralized, due to their large amount of SOEs, especially in defense industries. These provinces were also less active in initiating policies favorable to local development. Thus, local efforts, Mao's decentralization policy and the problems with socialist planning also contributed to the persistence of regional inequality in China.

Lastly, Mao's China was troubled by political and social turmoil: class struggle, rather than economic development, was considered the central task of the nation. While the First FYP did improve the economic status of the interior, this effort was soon disrupted. Except for the First FYP period, planning was never well made and fully implemented, and was interrupted by political and social conflicts. The GLF, Third Front, and Cultural Revolution wasted much of China's resources and had a devastating impact on China's economic development. National defense took much of China's resources, which otherwise could have been channeled to industrial production. The poor interior did not have the resources and the favorable macro conditions for its economic growth. In short, historical legacy, uneven geographic distribution of resources, the emphasis on industrialization and national defense, decentralization policies, and socio-political unrest have all contributed to the persistence of regional inequality in China.

Previous research tends to suggest that the governments succeeded or at least partially succeeded in achieving regional equality, as evidenced by the decline of interprovincial inequality in industrial production (e.g., Guo 1988). However, this research has found that, first, regional equality was only one of the factors influencing regional policies in the Maoist era: industrialization and national defense were often more important. Secondly, regional income inequality even rose under Mao, while regional industrial inequality was reduced only in a few years during the 1950s and persisted during the 1960s and 1970s. Thirdly, interior development policy, especially during the 1960s, was very selective, favoring a few provinces with strategic locations and excluding the poorest border and minority provinces. Moreover, interior investment was heavily concentrated in a few provincial capitals and industrial cities, while excluding most interior areas. As a result, intraprovincial inequality rose in many provinces, although further study is needed. Finally, due to heavy upward transfer and less investment, many coastal provinces lacked funds to improve urban infrastructure, to update industrial equipment, and to expand industrial production. Efforts placed on the interior and emphasis on national defense were at the cost of national industrial growth and efficiency, and kept the Chinese people poor. In short, policy under Mao was inconsistent and did little to improve regional development and the living conditions of the people. Problems with Mao's policy and the changing international context prompted the new leaders to adopt an open door policy and comparative advantage strategies.

4.6.2 Understanding post-Mao regional inequality

Patterns of uneven development during the reform period are more complicated than during Mao's era. Global forces, government policy, and regional endowments and efforts have all contributed to the change of regional inequality. First, the central state still plays an important role in China's regional development through initiating policies, although in comparison with Mao's era, the control and capacity of the central state have been declining, as evidenced by the declining ratio of budget revenue to GDP, the declining percentage of budgetary investment to total investment, the introduction of market mechanisms, and the declining share of industrial output produced by SOEs. The state has provided explicit regional development policies (such as the coastal development strategy) that are influential to regional development, and many non-regional policies are regionally uneven and are exerting a great effect on uneven development. The state has also provided favorable open door policies to the coastal provinces. Although some policies were implemented in the backward areas, they tend to be less effective. In short, the central state favors the growth of the coastal region and has contributed to the rise of interregional inequality and the rapid growth of a few coastal provinces.

Secondly, post-Mao economic reforms, including decentralization, open door policy and marketization, have increased the importance of regions to regional development. Scholars have long argued for the cellular nature of China's economy and the significance of local forces in China's regional development (Donnithorne 1972; Lyons 1985). Local governments actively initiate policies favorable to local development and tend to protect the interest of the localities. Local governments, especially those in the coastal region, have also actively participated in the national process of economic reforms and have intensely bargained with the central state to obtain favorable policies. Regionally different development paths and ownership structure are also critical to regional inequality. Non-state sectors have grown dramatically in several coastal provinces, including Guangdong, Zhejiang, Jiangsu, Shandong, and Fujian, while SOEs are more important in traditional industrial bases and many interior provinces. In addition, other local conditions, such as spatial linkages, local resource endowments, level of education and transportation, are also important to regional development.

Thirdly, globalization and China's open door policy have stimulated the growth of foreign investment in China, which also contributes to uneven development. As previous sections show, FDI has been concentrated in China's coastal region and has become an important source of investment in the coastal provinces. However, many interior provinces have attracted much less foreign investment. The coastal region has benefited more than the interior from the increase of FDI in China. The significance of FDI is further compounded by the fact that China's economy is still largely investment-driven.

Since economic reforms, the coastal region has outgrown central and western regions. But vast disparities exist among the coastal provinces, and traditional industrial bases in the coastal region have experienced slower growth. From 1978 to 1990, annual income and industrial growth rates of Shanghai, Beijing, Tianjin and Liaoning were among the lowest in the nation, mainly caused by slowing reform and opening up processes, the failure to revitalize SOEs and to renew their deteriorating industrial facilities and urban infrastructure. Their slow growth, especially Shanghai, the leading economic center of China, has greatly contributed to the decrease of interprovincial inequality. When socialist China was established in 1949, Shanghai was the most developed industrial center. Although the city's role in national industrial production declined during the 1950s, with a strong industrial base and a favorable state policy, it was able to maintain its leadership role during Mao's era. Economic reforms, however, severely eroded Shanghai's dominant economic role, especially during the 1980s. The opening up of many coastal cities to foreign investment, the price reform and marketization, and the drastic reduction of state-subsidized energy and raw materials greatly reduced the city's competitive edge. Meanwhile, the amount of the city's financial contribution to the state coffers and the strict control exerted by the central and municipal governments had not been reduced. The old and overburdened urban infrastructure was also deteriorating. As a result, the dominant role of Shanghai as a leading producer declined, leading to the corresponding decline of interprovincial inequality in China.

Although the dominance of the eastern region remains, the ranks of provinces in the national economy have changed. By 1990, Jiangsu, Shandong, and Guangdong became the biggest producers of industrial output, while Shanghai fell to fourth place (or 10th place in terms of national income). The coastal provinces of Jiangsu, Guangdong, Shandong, Zhejiang, and Fujian have grown more rapidly during the reform period. They have acquired more favorable policies in fiscal systems, investment allocation, foreign investment and trade, ownership transformation, administrative power, and many other areas. They have initiated numerous programs to stimulate economic growth, and have advantages in factor endowments. They have also dominated foreign investment and trade. Their rapid growth, has contributed to the more rapid growth of the coastal region as a whole (the rise of interregional inequality) and the declining status of the traditional industrial bases (the decline of interprovincial inequality).

I have performed a regression analysis to reveal the determinants of uneven regional growth, with a focus on investment.[13] The dependent variable is average annual growth rate of per capita national income of the provinces (NIPG) for the entire 1978–92 period. Growth rate is the most widely used dependent variable in statistical analyses of the underlying forces of regional inequality (e.g., Chen and Fleisher 1996). The unit of analysis is China's provinces. The period of analysis is from 1978, the year when economic reforms were launched, to 1992, the most recent year for which national income data for China's provinces were available (GDP can also be used for the regression analysis).

I have selected four independent variables (Table 4.18):

1 Investment growth is critical to economic growth in the socialist countries, and has significantly contributed to uneven regional growth in China. Regional investment growth is expected to positively contribute to regional growth. The data start in 1982, as it was the first year when SSB reported regional data on fixed asset investment (FAI) in its statistical yearbooks.
2 It has been well documented that foreign investment tends to locate in core regions and intensify regional inequality in developing countries. FDI has been widely used to represent the importance of global forces, especially foreign investment, and should positively contribute to regional growth.
3 Ownership structure reflects institutional changes, local development paths, local initiatives and local resource endowments, although state policy and foreign investment also exert some influence. Many coastal provinces have invested heavily in non-state sectors that are expanding rapidly, while lagging interior provinces are more dependent on SOEs. SOE is used to represent regional ownership structure and is expected to negatively relate to regional NIP growth.
4 A COAST dummy variable is assigned to the five coastal provinces as they have advantages in local conditions that may include local governments, access to international marketplaces and resource endowments. It is a variable used to show effects of local forces on regional inequality, besides the consideration of FAI, FDI, and SOE.[14] This variable is particularly important to China as substantial coastal and interior disparity has long existed.

The results of regression analysis are shown in Table 4.18. The model is tested for regression assumptions and no violation was found. The regression equation as a whole and the regression coefficients are all significant at either

Table 4.18 Results of regression analysis

Independent variable	*Standard coefficient*	*Error*	*Standard regression coefficient*	*T-value*	*Significance level (P)*
FAI	0.004	0.001	0.382	0.374	0.0069
SOE	−0.025	0.015	−0.191	−1.739	0.0954
FDI	0.396	0.223	0.160	1.773	0.0894
COAST	1.669	0.652	0.360	2.560	0.0175
Constant	6.969	1.352	0.0	5.154	0.0000

Dependent variable NIPG
$R = 0.916$ $R^2 = 0.858$ Adjusted $R^2 = 0.833$
$F = 34.655$ Significance level $(P) = 0.0000$

the 5 percent or 10 percent level. Multiple R^2 is as high as 0.858, indicating that 85.8 percent of the variation in regional growth rate is explained by the independent variables. In other words, only 14.2 percent of the variation in income growth is not explained.[15] The high R^2 indicates that investment growth, foreign investment, ownership structure, and location are significant factors underlying regional inequality in China. Regression analysis shows that NIPG is positively related to FAI, FDI, and COAST, indicating that provinces with faster investment growth, with larger shares of FDI, and coastal in location, tend to grow faster. NIPG is negatively related to SOE, indicating that provinces with more significant presence of SOEs tend to lag behind. In terms of the relative importance of the independent variables, as indicated by standard regression coefficients, FAI is the most important, followed by COAST, SOE, and FDI. Thus, it can be generally concluded that both state policy and local forces play important roles in regional growth, followed by foreign investment. Regression analysis strongly supports my argument for the importance of state policy, regions/localities, and foreign investment in China's regional development.

In short, the state, regions and foreign investment are three influential forces driving China's uneven development. The central state is no longer the dominating force underlying regional inequality in China. The region has emerged as another important agent in shaping China's regional development. In addition, foreign investment is concentrated in China's coastal provinces and has stimulated the growth of the coastal provinces; i.e., regional inequality in China can be understood by considering the state, regions and global forces.

4.7 Conclusion

Issues of equality and inequality have been important to Chinese leaders since the establishment of the PRC. Some efforts were made by the central government in allocating state investment in a spatially more balanced way to reduce regional inequality. Since the launch of reforms, rapid growth and profound changes have taken place in China. Uneven development was considered necessary to achieve rapid and efficient economic growth. A spatially balanced pattern of development is no longer emphasized, but the idea has never been totally discarded and has been re-emphasized lately.

When the PRC was established, China's regional inequality was substantial. Although efforts were made by Mao to reduce the coastal–interior disparity, regional inequality persisted during the 1952–78 period. In the Maoist era, interior development was not successful as it did not significantly stimulate interior development and reduce regional inequality. Poor interior investment and local efforts have seriously hindered the achievement of interior development. Given the large differentials in historical legacy and local context, the disparity between China's coastal and interior regions was difficult to eliminate.

Post-Mao reforms have promoted decentralization, marketization, and globalization. The control and capacity of the central state have declined, and global and local forces have emerged as important forces underlying China's regional development. During the reform period, the coastal provinces in general have received more preferential policies than the interior. Local governments in the coastal provinces have also actively initiated policies favorable to local development. Moreover, foreign investment and trade are also concentrated in a few coastal provinces. As a result, interregional inequality has risen further, and its increase was accelerated during the early 1990s. Overall, the core–periphery structure mainly established during China's semi-colonial period has persisted after four and a half decades of socialism.

Interprovincial inequality, however, declined. This is largely due to the slow growth of the four traditionally leading provinces (Shanghai, Beijing, Tianjin, and Liaoning) and the rapid growth of the five emerging coastal provinces (Guangdong, Jiangsu, Zhejiang, Shandong, and Fujian). Post-Mao economic reforms have dramatically restructured China's regional economies and a new form of uneven regional development has emerged. Regions can no longer rely on the policies of the central state for their economic development. Rather, local conditions and foreign investors are exerting a great impact on regional development; this seriously challenges the recent drive of the central government to reduce poverty and regional inequality in China.

5 Patterns of regional development in Jiangsu

While some progress has been made in the understanding of regional development across China's macro regions and provinces, much less attention has been paid to regional development and inequality at the subprovincial level. The emphasis of research on interregional and interprovincial scales is mainly because they are the focus of central government policy and data are readily available for analysis. Intraprovincial inequality, on the other hand, is largely an issue of local governments, and can hardly become a national agenda. Moreover, data constraints have limited the research. However, this level of analysis can reveal detailed processes and forces that operate locally. In addition, substantial regional inequalities exist within the provinces, particularly in those experiencing more rapid growth.

To provide a more detailed analysis of regional development in provincial China, I have chosen Jiangsu as a case study. I analyze multiple mechanisms and factors underlying regional development in Jiangsu, including central government policies, local initiatives, and global forces, especially through an analysis of fiscal systems, fixed investment, foreign investment, and industrial ownership structure. I examine three geographical scales of regional inequality in Jiangsu—interregional, overall intercounty, and rural intercounty inequalities—with an emphasis on interregional and intercounty inequalities. In this book, I use interregional inequality to refer to inequality across the three macro regions, i.e. southern Jiangsu (Sunan), central Jiangsu (Suzhong), and northern Jiangsu (Subei); overall intercounty inequality refers to inequality across all county-level administrative units (including counties, county-level cities, and city districts of municipalities/prefectures); and rural intercounty inequality refers to inequality across counties (including counties and county-level cities).[1] For simplification, I refer to overall intercounty inequality as intercounty inequality and rural intercounty inequality as rural inequality. While interregional inequality across Sunan, Suzhong, and Subei is the major concern of the provincial government of Jiangsu, intercounty and rural inequalities are also important dimensions of intraprovincial inequality and should be investigated.

5.1 The Jiangsu case and research method

As I have discussed in previous chapters, although Mao made some efforts in developing China's backward regions, regional inequality under Mao persisted. During the reform period, patterns of regional development are more complex than often assumed. Regional development at the subprovincial level has been examined less, despite the fact that considerable differences exist within the provinces and the uneven impact of economic reforms on different cities and regions merits more research (Goodman and Segal 1994; Solinger 1996; Wei and Fan 1997).

Although a few recent studies have investigated intercounty inequality in the reform period, findings are mixed, partly due to the lack of data and partly because of different research methods. For example, Fan (1995a, 1995b) found that in the 1982–90 period, intercounty inequality surged in Guangdong, remained about the same in Fujian, and declined in Zhejiang, Anhui, Hunan, and Jiangsu. Rozelle (1994) found a trend of increasing intercounty inequality in Jiangsu, and Weng (1998) showed a decline of regional disparity in Guangdong during the reform period.[2] Using CVs for per capita rural gross social product, Yang (1997) and H. Wei *et al.* (1997) found intercounty inequality increased significantly between 1985 and 1990 in most provinces. Lyons (1998) also found a rise of intercounty inequality in Fujian, although every county in Fujian has experienced substantial growth and alleviation of poverty.

Studies based on rural income inequality and intervillage inequality, which are only marginally relevant to this research, maintain that rural inequality under Mao was at a low level (World Bank 1990; Bramall 1993), although some scholars argue that intervillage inequality actually widened (Friedman *et al.* 1991). During the reform period, while some scholars maintain rural inequality in China was fairly modest (World Bank 1992b), others argue intervillage inequality has been intensified (Bramall and Jones 1993; Rozelle 1994, 1996). In addition, scholars have observed that rural reforms did not intensify rural inequality (World Bank 1985), but comprehensive urban and market reforms have intensified it (e.g., Lo 1990; Bramall and Jones 1993; Knight and Song 1993).

Previous scholarly efforts have certainly enriched our understanding of changing patterns of regional development within China's provinces. However, much of the research on intraprovincial inequality is based on limited data, rather than systematic work based on time series county-level data. More efforts are needed to reveal a full picture of intraprovincial inequality. Moreover, previous studies tend to focus on the measurement of regional inequality, and have paid less attention to revealing the causal factors underlying regional development and inequality. Nevertheless, economic reforms have produced a complex landscape of regional development at the subprovincial level. An analysis of disaggregated county-level data should allow me to reveal causal factors underlying the change of regional

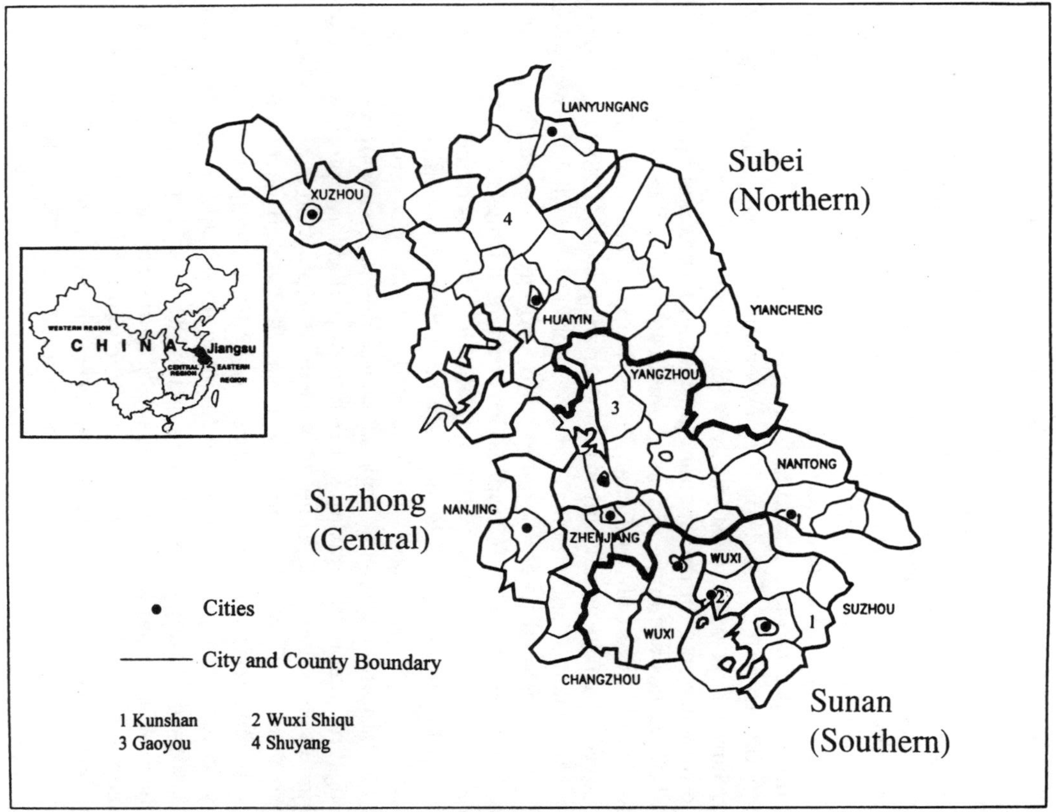

Figure 5.1 Regions and counties in Jiangsu.

inequality. A detailed analysis of Jiangsu should shed more light on the understanding of intraprovincial inequality in China both under Mao and in the reform period.

Jiangsu is selected as a case study of intraprovincial inequality mainly for four reasons. First, it has recorded rapid growth, and its growth trajectory reflects that of many other coastal provinces, such as Zhejiang and Shandong. From 1978 to 1995, its average annual growth rate of per capita GVIAO was 13.5 percent, among the highest in China. In 1995, its per capita GVIAO was 11,129 yuan, also among the highest. Jiangsu has been one of the most dynamic provinces of China and has always been in the forefront of national policy changes. The province has been sensitive to state policy change, because it is one of the most developed provinces and because of its geographic proximity and close ties to Shanghai, which has long been an epicenter of political and economic changes in China.

Secondly, large regional disparities, especially between poorer Subei and richer Sunan, have existed historically. While a notion of a three-fold division of the province into Sunan, Suzhong and Subei has gained favor (Fig. 5.1), spatial disparities within the province remain substantial.[3] In 1995, per capita GVIAO in Subei was only 4667 yuan, far less than that in Sunan (25,983 yuan). The gap is even more astonishing in terms of per capita FDI (US$9 vs. US$259). Rising regional gaps and conflicts have prompted the provincial government to consider 'regional co-development' a top priority of the Ninth Five-Year Plan of Jiangsu (1996–2000) (People's Government of Jiangsu Province 1997).[4] However, to what extent regional inequality has increased and what forces have contributed to the change have not been thoroughly studied.

Thirdly, studies on Jiangsu remain limited and have not revealed the complex mechanisms underlying regional development. A few recent studies investigating regional inequality in the reform period have produced mixed findings (Veeck 1991b; Rozelle 1994; Fan 1995b). Moreover, studies on Jiangsu tend to emphasize the importance of TVEs in development (Ho 1994). While I recognize the importance of TVEs, I attempt to analyze state policy, local factors, and foreign investment to understand the development of TVEs and the change of regional inequality. The fact that regional inequality differs with geographic scales needs to be studied more thoroughly. In this and the following chapter, I examine the functioning of multiple mechanisms in regional development in Jiangsu, including global, state and local forces, and demonstrate the importance of geographic scales in studying regional inequality.

Finally, availability of subprovincial data provides another incentive for me to study Jiangsu. The Jiangsu Statistics Bureau (JSB) has published detailed county-level statistics that are hardly matched by any other province (JSB 1989–1997, 1994b). As well as analyzing this data, I synthesize information gathered from my years of research and experiences with Jiangsu which began in the mid-1980s, and informal interviews with local officials and scholars, to help data interpretation.

Jiangsu implemented a 'city-leading-county', system (*shi dai xian*) in 1983.[5] Under this new system, the governments of prefecture-level cities administer surrounding counties and county-level cities. After 1985, Jiangsu was divided into 11 prefectural-level municipalities (or prefectures for consistency) that include city districts (*shiqu*), counties and county-level cities.[6] In this research, three levels of regional inequality in Jiangsu will be analyzed. First, the inequality across Sunan, Suzhong and Subei (interregional inequality) has long existed and has generated considerable scholarly and governmental concerns. Secondly, I analyze regional inequality in Jiangsu based on all county-level units (counties, county-level cities, and city districts)—the basic building blocks of China which is called intercounty inequality. Thirdly, although not emphasized, I examine inequality across counties and county-level cities in Jiangsu, which is referred to as rural inequality.[7] A set of time-series data (from 1950 to 1995), which is internally consistent in definitions and administrative boundaries, is used for documenting and analyzing changes of regional inequalities in Jiangsu.

The major indicator used is GVIAO per capita which has often been employed as an effective indicator of the Chinese economy, particularly for the county level.[8] To effectively compare changes over time, data for comparable prices (real GVIAO), calculated from growth indexes will be used. Since JSB did not release the growth index for 1993, I averaged the inequality indices for 1992 and 1994 to estimate their respective levels for 1993. I adopt the coefficient of variation (CV) and the location quotient (LQ) for their simplicity and popular usage. I also use regression analysis, which will be discussed in the relevant section.

5.2 Development process of Jiangsu

5.2.1 The making of the Sunan–Subei divide

Jiangsu has long been one of the most developed provinces of China. The oldest farming systems and human settlements were found in both Subei and Sunan. Sunan in particular is part of the Yangtze Delta Plain which enjoys a favorable setting (e.g., abundant rainfall, fertile soil, and well-established water irrigation systems) for fisheries and agriculture, and is often referred to as the 'land of fish and rice' (*yumi zhixiang*). The city of Suzhou became the capital of the Prince of Wu around 500 BC, and emerged as a military and economic center for a vast surrounding area. During the Han dynasty, the development of the salt industry and the construction of canals stimulated the growth of Subei, especially cities such as Yangzhou and Huaiyin. During the period of disunity (220–581), Subei, especially the Huaiyin and Xuzhou regions, were involved in wars, while Sunan had a relatively stable condition for economic development.

In the Tang dynasty (618–907), Jiangsu developed into one of China's most prosperous provinces, and was linked to the national capital in the north through the Grand Canal for tribute grain (*cao liang*) transport from south China to north China. Cities along the Grand Canal, whether in Sunan or Subei, were prosperous transport and trading centers. Indeed, Yangzhou and Huaiyin, as centers for grain transport along the Grand Canal and both located in Subei, were among China's most prosperous cities. During the Song dynasty (907–1279), the development of irrigation systems, transportation, and technology greatly stimulated agricultural development in Sunan. Handicraft industry and trading activities boomed in Jiangsu, which, along with Zhejiang, was critical to the filling of the state coffers. However, after the Yellow River joined the Huai River in 1128 (in the Song dynasty), Subei was persistently troubled by serious flooding for the following centuries (Zhang and Su 1993). It had less favorable conditions for agricultural development, and became a battlefield, especially during the Song dynasty, contributing greatly to the rising gap between Sunan and Subei.

During the Ming dynasty (1368–1644) and the early Qing dynasty (1644–1840), with the development of the commercial handicraft industry, especially cotton and silk production (Zhang and Su 1993), Sunan was the wealthiest region of China. Hundreds of townships functioned as centers of commercial activities and linked the cities to the rural economy (Marme 1993). The population also increased rapidly in Sunan, and in 1819, Jiangsu was the most populous province of China. In Subei, however, from 1400 to 1900, flooding broke out 350 times, which seriously damaged its agricultural and living conditions, and led to the stagnation of Subei (Editorial Committee 1989). Especially from the mid-1800s onward, cycles of flooding, famine, and poverty characterized large portions of the region. It became the embodiment of poverty and rural backwardness, and over the years, Subei has differentiated itself from Sunan in geography, language and culture (Honig 1992).

The relative status of Subei declined further with the changing mode of transportation. In 1282 (in the Yuan dynasty), sea transport began to replace the Grand Canal for the shipment of grain from south China to north China (Zhang and Su 1993). Sea transport became the major means of tribute grain transport in the 1840s. Cities in Subei, which previously functioned as centers of transportation and commercial activities, gradually lost their economic bases and declined further. The deterioration was intensified by the shift of the course of the Yellow River in 1853, and the Huai River once again became a course of treacherous flooding in Subei (Honig 1992).

Subei's decline contrasted sharply with the rise of Sunan. First, the Huning Railway linking Nanjing and Sunan with Shanghai was built in 1908, and was connected to north China (Tianjin) in 1911. Transportation conditions in Sunan were dramatically improved, while vast Subei lacked modern transportation. Meanwhile, Shanghai gradually emerged as a major economic center in Pacific Asia. Sunan, through the development of modern transportation and communications, was well integrated into the networks of cities and market places in China.

Secondly, with the foreign invasion after the Opium War and especially the Sino–Japanese War in 1895, foreign investors and Chinese nationalists began to build modern industries in Jiangsu, including food processing, textile, and manufacturing industries. At the beginning of the 20th century, Jiangsu had established 155 modern enterprises, and by 1932, Jiangsu had 4652 modern enterprises that employed 108,000 workers (Editorial Committee 1989). Most of these enterprises were located in Sunan, particularly in Suzhou, Wuxi, and Changzhou, which were among the most important industrial cities in China at that time. But Subei was still troubled by flooding,[9] and its status in the economy of Jiangsu declined further.

Lastly, Sunan's industrial development also benefited from the improvement in agriculture and the rural economy. In the rural areas, commercial agriculture and handicraft production boomed, and small towns became centers of rural economies. Cotton and silk cultivation provided raw materials for the growth of textile and silk industries in Sunan, making Suzhou, Wuxi, and Changzhou China's major industrial centers. These cities were linked to rural areas and small towns in Sunan. On the other hand, booming rural areas and small towns provided market places for industries in the cities.

All these changes produced and intensified the development gap between Sunan and Subei. Honig (1992) described the contrast between Sunan and Subei as follows:

> ...one is struck by accounts of the wealth of the south versus the poverty of the north; the lush scenery of the south versus the drab, dreary north; the soft Wu dialect spoken in the south versus the harsh Yangzhou dialect of the north; the industrial economy of the south versus the agricultural economy of the north; the fertile soil of the south versus the dry, alkali soil of the north.
>
> (Honig 1992, p. 28)

5.2.2 Jiangsu under Mao

Before the establishment of the PRC in 1949, Jiangsu had already established numerous modern enterprises and was one of the most developed provinces of China. In 1952, per capita GDP and per capita GVIAO were 131 yuan and 155 yuan, respectively, among the highest in China (Table 5.1). In 1950, with 1 percent of China's total land area, Jiangsu produced 9.9 percent of China's industrial output, although its per capita NI (or provincial income) and industrial output was only at the national average. Light industry, especially small-scale textiles and food processing, was particularly important to Jiangsu. Its rural areas were also relatively more developed than many other rural areas of China.

During the early and mid-1950s, Jiangsu's private enterprises were gradually transformed into either state-owned enterprises (SOEs) or collective-owned enterprises (COEs). Most of the small-scale private enterprises were

Table 5.1 Major economic indicators in Jiangsu, 1952–95

Economic indicator	*1952*	*1978*	*1995*	*1952–78 average annual growth (percent)*	*1978–95 average annual growth (percent)*
Population (million)	37.4	58.3	70.7	1.7	1.1
GDP (billion yuan)	4.8	24.9	515.5	5.2	12.8
Per capita GDP (yuan)	131	430	7,299	3.4	11.5
GVIAO (billion yuan)	5.7	44.4	1,431	8.9	19.6
Per capita GVIAO (yuan)	155	765	20,317	7.0	18.3
Fixed (asset) investment (billion yuan)	0.12	2.18	168.02	11.8	29.1
Foreign investment (US$ billion)	—	—	5.29	—	—
Export (US$ billion)	—	0.42	11.79	—	21.7

Source: JSB 1996.
Note: Current prices; growth rates in comparable prices (except investment and export).

Table 5.2 Industrial output in Jiangsu, 1950–95 (billion yuan)

		SOEs		*COEs*		*VEs*		*Other*	
Year	*Total*	*Volume*	*Share*	*Volume*	*Share*	*Volume*	*Share*	*Volume*	*Share*
1950	1.9	1.2	63.2	0.7	36.8	0	0.0	0.0	0.0
1960	10	8.2	82.0	1.8	18.0	0	0.0	0.0	0.0
1970	13.5	10.8	80.0	2.5	18.5	0.3	1.5	0.0	0.0
1975	23.5	16.2	68.9	6.5	27.7	0.8	3.4	0.0	0.0
1978	33.8	20.8	61.5	10.5	31.1	2.5	7.4	0.0	0.0
1985	103.7	42.6	41.1	41.6	40.1	17.4	16.8	2.1	2.0
1990	276.4	94.8	34.3	109.5	39.6	61.0	22.1	11.1	4.0
1995	1,262.3	249.8	19.8	478.8	37.9	360.6	28.6	173.1	13.7

Source: JSB 1996.

transformed into COEs, which have since become an important sector in Jiangsu. The number of SOEs only accounted for about 3 percent of industrial enterprises, much lower than the national average. Due to its coastal location and lack of energy and raw materials for heavy industry, fewer SOEs were newly built in Jiangsu during Mao's era. In 1960, SOEs produced 82 percent of industrial output in Jiangsu, which was still lower than the national average (91 percent) (Table 5.2). Because of this, small-scale enterprises and COEs have been more important in Jiangsu than in many other provinces.

China's investment policy in the 1950s favored the recovery of traditional industrial bases and the development of the interior region, while most of the coastal provinces, including Jiangsu, were de-emphasized in resource allocation. During the First FYP period, among the 156 key-point projects imported from the Soviet Union, none was allocated to Jiangsu. However, Jiangsu still had to contribute substantial revenues to the state. Moreover, some industrial equipment and technicians in Jiangsu were transferred to poorer interior provinces for the purposes of national defense and interior development. Within Jiangsu, much of the investment during the First FYP period went to renewing the established industries, which took in 386, or 93.4 percent, of the 413 large investment projects in Jiangsu. Most of these projects were based in Sunan, although some small plants were established in Subei for processing agricultural products (Editorial Committee 1989).

Jiangsu's economic growth was among the lowest in the nation. Growth rates of per capita national (or provincial) income were even negative during most of the 1950s, as a result of slow economic growth and rapid population growth (Fig. 5.2). As shown in Fig. 5.3, its relative status, as measured by LQs of per capita NI, declined substantially during the 1950s. In 1952, the LQ for Jiangsu was 97 (near average), which declined to 79 in 1957.

During the Great Leap Forward (GLF) period (1958–60), while Jiangsu experienced chaos and economic crisis, decentralization did give Jiangsu some decision-making powers in developing local economies. After Mao's calling for rural industrialization, some small-scale rural industries were established, particularly in Sunan: most of them were food processing and agricultural machinery factories. However, the failure of the GLF forced the closure of numerous enterprises, and millions of urban workers were sent back to the countryside. In 1961, the government closed or rectified 5238 industrial enterprises, and the numbers of COEs and SOEs were reduced 50 percent and 40 percent, respectively (Editorial Committee 1989). The LQ of

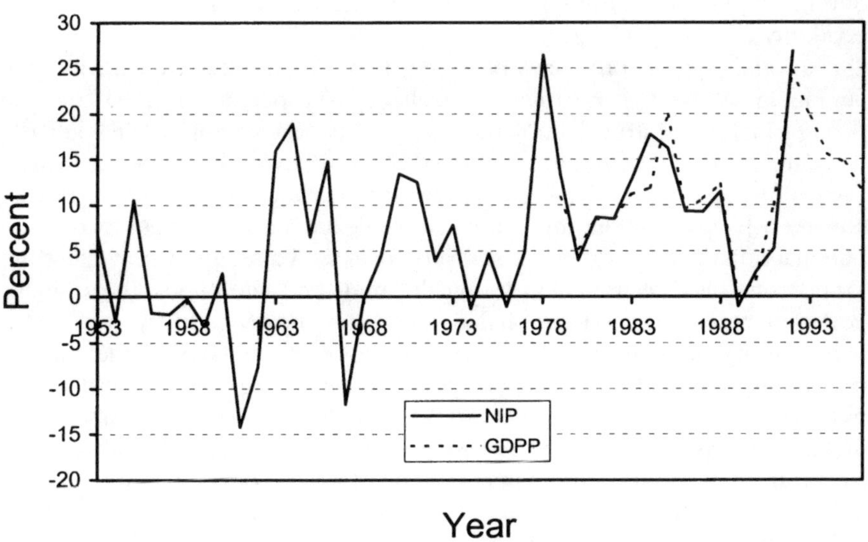

Figure 5.2 Growth rates of national income per capita (NIP) and GDP per capita (GDPP), 1953–95.

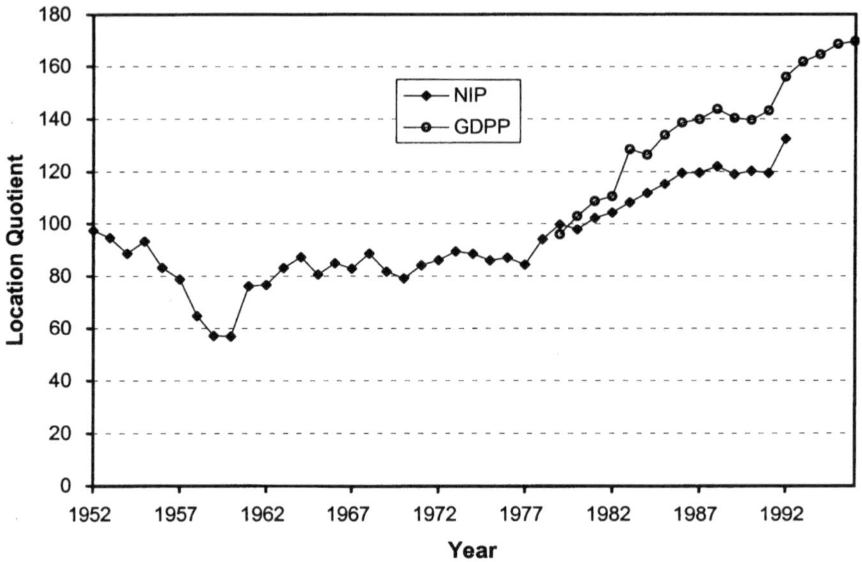

Figure 5.3 Location quotients based on national income per capita (NIP) and GDP per capita (GDPP), 1952–95.

Jiangsu decreased further, from 79 in 1957 to 57 in 1959–60, the lowest level during Jiangsu's post-revolution development. The change of LQ in Jiangsu during the 1950s indicates a dramatic decline of its status in the national economy.

Like other coastal provinces of China, Jiangsu experienced a devastating famine in 1961–62. Per capita NI declined 14.2 percent in 1961 and 7.6 percent in 1962. With the implementation of economic adjustment policies, Jiangsu's economy began to recover in 1963, followed by stable growth in 1964 and 1965 (Fig. 5.2). However, the Cultural Revolution seriously damaged Jiangsu's economy, although the leadership in Jiangsu was more pragmatic than many other coastal provinces and attempted to protect the local economies. Meanwhile, during the mid- and late 1960s, the Chinese government emphasized national defense, and heavy industry received a high priority in state investment. As the government favored the interior region, especially the Third Front area, in allocating defense and heavy industries, state investment in Jiangsu was very limited. Moreover, investment in Jiangsu was biased towards heavy industry as well, although Jiangsu lacked the energy and raw materials needed for the development of heavy industry. The state sector was also favored in development, while collective industries were discriminated against, and private enterprises were almost totally eliminated. Jiangsu recorded negative growth rates in 1967–68, and its status in the national economy did not improve (Figs. 5.2 and 5.3).

With the normalization of Sino–US relations and the emergence of a pragmatic leadership in the early 1970s, the emphasis of China's regional development began to shift to the coastal region. Several large-scale investment projects were allocated in Jiangsu, particularly in Nanjing. Meanwhile, the central government encouraged the development of small-scale rural enterprises, followed by the establishment of hundreds of TVEs in Jiangsu, particularly in Sunan. From 1970 to 1975, rural industrial output increased from 0.96 billion yuan to 2.33 billion yuan, and industrial output surpassed agricultural output for the first time. In 1975, SOEs produced 68.9 percent of Jiangsu's industrial output, which was less than the national average (81 percent) (Table 5.2). The growth of TVEs contributed greatly to the growth of Jiangsu's economy in the early 1970s (Fig. 5.2). The status of Jiangsu in the national economy began to recover, as shown by the slight rise of LQ in Fig. 5.3.

During Mao's era, growth rates of major economic indicators were low compared with the post-Mao period (Table 5.1), and the growth fluctuated greatly over time. From 1952 to 1978, the growth rates of per capita NI and per capita GDP in Jiangsu were 3.8 percent and 3.4 percent, respectively, lower than the national average, although the growth rate of per capita GVIAO was higher. Industry recorded more rapid growth than agriculture, and heavy industry grew more rapidly than light industry. Manufacturing, textile, chemical, and food industries were the major sectors. In 1977, the LQ of Jiangsu based on per capita NI was 84, still lower than its level in the early 1950s and the national average (100), indicating the decline of Jiangsu's relative status in the Chinese economy during Mao's era. The economic structure of Jiangsu to a certain extent resembled that of China, such as the dominance of SOEs, the emphasis on heavy industry and the low efficiency of investment, although Jiangsu had more collective enterprises and light industries than many other provinces of China.

5.2.3 Jiangsu under economic reform

After the initiation of reforms in 1978, the provincial government began to restructure the economy of Jiangsu, including developing light industry and collective enterprises and improving industrial and agricultural productivity, in an attempt to solve some problems associated with state socialism. Jiangsu negotiated a contract fiscal system with the central government in 1977, ahead of other provinces. Rural reforms were introduced in the late 1970s in Jiangsu, and various contract systems were implemented in rural areas. The effects of such reform policies were evident by the double-digit growth rates of per capita NI (26.4 percent in 1978 and 13.2 percent in 1979) (Fig. 5.2). By the summer of 1983, 98.6 percent of villages implemented the rural responsibility system (Editorial Committee 1989). Rural reforms stimulated agricultural development, which in turn provided capital, raw materials, and labor for rural industrialization. The city-leading-county system was also introduced in that year to improve linkages and production cooperation between cities and the countryside. Since the late 1970s, Jiangsu's relative status in the Chinese economy has been steadily rising (Fig. 5.3).

In 1984 and 1985, the emphasis of reforms was shifted to more comprehensive urban and industrial reforms, including the planning system reform, industrial management reform, fiscal reform, investment reform, price reform, wage reform, and commercial system reform. Like other coastal provinces, Jiangsu received more decentralized and more open policies from the central government. Fiscal and investment reforms have provided great incentives for the growth of local economies, and administrative reforms have decentralized many SOEs from central and provincial governments for local management. Although opened up to the outside world later than Guangdong and Fujian, Jiangsu has been vigorous in attracting foreign investment and promoting international trade, and has captured the interest of foreign investors. In 1996, FDI in Jiangsu reached US$5.2 billion, only behind Guangdong. Reforms have also been carried out in rural enterprises, with a focus on management and ownership structure of TVEs.

Post-Mao economic reforms have brought rapid growth and profound changes in Jiangsu. During the reform period, major economic indicators recorded double-digit growth, among the highest in the nation, although growth rates of per capita NI and GDP were negative in 1989 due to the Tiananmen incident and policy changes. As shown in Table 5.1, GVIAO per capita increased from 765 yuan in 1978 to 20,317 yuan in 1995, an increase of 11.5 percent annually. Rural industrial output increased from 6.24 billion yuan in 1978 to 778.9 billion yuan in 1995, an annual growth of 31 percent, much higher than the growth rate of agricultural output (7 percent annually), and output of TVEs alone increased from 6.23 billion yuan to 706 billion yuan. Economic growth in Jiangsu was accelerated in the early and mid-1990s. Between 1990 and 1995, industrial output increased annually at 31.5 percent, and rural industrial output increased annually at 42.3 percent.

Jiangsu's status in the national economy has been improved dramatically. Its growth rate of per capita GDP during the 1978–95 period was 11.3 percent, ranked third among China's provinces (only behind Zhejiang and Guangdong). LQs based on per capita GDP increased from 119 in 1978 to 170 in 1995, an increase of 42.8 percent, contributing greatly to the rising status of the coastal region in the Chinese economy (Fig. 5.3). Its fixed asset investment, as a percentage of China's total, increased from 6.1 percent in 1982 to 9.3 percent in 1995. Its FDI, as a percentage of China's total, increased from 1.5 percent in 1983 to 13.9 percent in 1995. In 1995, its GDP reached 515 billion yuan, only slightly less than Guangdong (538 billion yuan). With 1 percent of the total land area of China and 5.8 percent of China's population, Jiangsu produced 8.9 percent of GDP and 12.9 percent of industrial output (GVIO) in China. Jiangsu's economy is also more open and better managed than many other provinces. In 1995, Jiangsu's exports accounted for 11 percent of the total exports of China. The economy of

Jiangsu is much less controlled by the central government and state planning. In 1995, as shown in Table 5.2, SOEs only produced 19.8 percent of GVIO, much lower than the national average (34 percent). Investment is also largely beyond state control, as investment from state budget and bank credits only accounted for 20.4 percent of CCISOE.

Reforms have provided favorable conditions, especially for the development of non-state enterprises. As shown in Table 5.2, SOEs have recorded slower growth. The share of industrial output produced by SOEs has declined yearly. As Jiangsu's enterprises tend to be smaller in scale and it has traditionally had lower proportions of output produced by SOEs, Jiangsu's economy has been less hindered by the slow growth of SOEs. TVEs in particular have experienced dramatic growth and have become the dominant sector in the economy. TVEs have blossomed in Sunan, well known to the nation as the Sunan model. The development of private enterprises has also been accelerated in the 1990s, especially in the service sector. Meanwhile, the province encouraged Subei to develop different models of economic growth based on local comparative advantages.[10]

Moreover, considerable efforts have been made to improve interprovincial and intraprovincial transportation. Jiangsu has established one of the most efficient transport systems in China, consisting of airlines, ports, railways, waterways, and highways, but Subei has poorer transportation conditions. Major social and development indicators, such as income, savings, rate of tap water use, school enrollment, and health care facilities, are all among the highest in China (Table 5.3). Indeed, Jiangsu's social and infrastructure development has benefited greatly from economic growth and the efforts of local leaders in improving the living conditions.[11]

However, Jiangsu's impressive economic growth was not free of problems. Its population density of 689 persons per square kilometer is much higher than the national average (126 persons/km^2) and is one of the highest among provinces of China. Economic growth and urbanization have placed a higher pressure in preserving resources and protecting the fragile environment. TVEs have been troubled by the problem of poorly defined property rights and poor economies of scale and agglomeration. The province has always handed considerable revenues to the central government, but Jiangsu's poor

Table 5.3 Selected development indicators in Jiangsu, 1995

	Jiangsu	*China*
Per capita net income of rural residents (yuan)	2,457	1,578
Savings per capita (yuan)	2,721	2,449
Medical professional (per 10,000)	35	10
College students (per 10,000)	30	9
Rate of household gas use (percent)	81.8	70
Rate of tap water use in urban areas (percent)	100	93

Source: SSB 1996; JSB 1996.

regions benefit little from these transfers. Interregional conflicts and regional problems have become serious, and the issue of regional inequality has concerned many scholars and local governments.

5.3 Trends of regional inequality

5.3.1 Interregional inequality across Sunan, Suzhong, and Subei

For government officials in Jiangsu, interregional inequality across Sunan, Suzhong, and Subei is the most important dimension of regional inequalities in Jiangsu. It has generated considerable concerns, especially for government officials, scholars, and people living in poorer Subei. Meanwhile, cities and counties (including county-level cities and city districts) are the building blocks of the lower-level administrative divisions of China. Regional development across those lower administrative levels has also attracted considerable scholarly attention. In this book, I analyze regional inequalities across regions, cities and counties, and counties in Jiangsu, which are referred to as interregional, intercounty, and rural inequalities for simplicity.

As I have discussed previously, Sunan had long been one of China's most prosperous agricultural and commercial bases. It also established numerous modern enterprises during the early 20th century. Wuxi and Suzhou were among the most industrialized cities in China. Subei, however, fell behind. With the establishment of the PRC in 1949, new policies were implemented to develop Jiangsu's economy. Figure 5.4 shows that interregional inequality

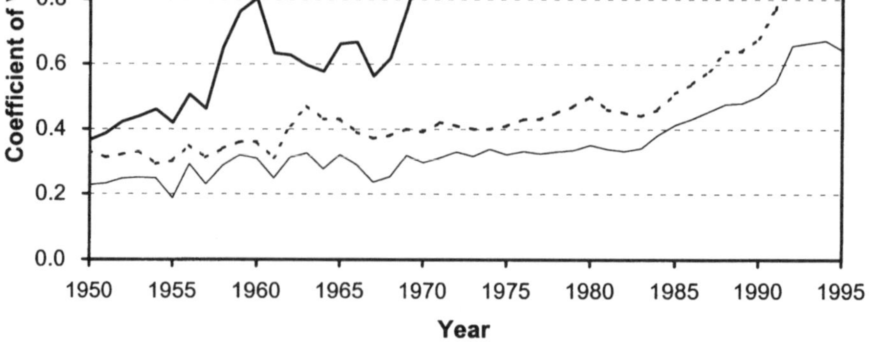

Figure 5.4 Trends of interregional, intercounty, and rural inequalities in Jiangsu.

among Sunan, Suzhong, and Subei increased during Mao's era, contrasting with the widely held notion that regional inequality declined under Mao. The CVs based on per capita GVIAO rose from 0.23 in 1950 to 0.33 in 1978, an increase of 43.5 percent.

Interregional inequality only increased slightly during the early and mid-1950s and much of the 1960s when considerable resources were committed to the development of China's poorer interior regions. But it was widened especially during the late 1950s and early 1970s. These two periods corresponded with China's decentralization, during which the provincial and local governments of Jiangsu received some decision-making power from the center. In Subei, agricultural productivity was improved by the advancement of irrigation and agricultural technology, as evidenced by a faster growth of agricultural output in Subei, although some new industries were also established. Sunan, however, through its industrial growth and a larger industrial sector, still developed faster than Subei. Sunan's growth rate of per capita GVIAO during 1950–78 was 5.4 percent, faster than Subei (4.3 percent) and Suzhong (5.2 percent) (Table 5.4). As a result, interregional inequality in Jiangsu increased during Mao's era.

During the post-Mao period, interregional inequality in Jiangsu increased rapidly, as shown in Figs 5.4 and 5.5. During the late 1970s and early 1980s when rural reforms were emphasized, interregional inequality increased only

Table 5.4 Per capita gross value of industrial and agricultural output, 1950–95 (yuan)

Region	*1950*	*1978*	*1995*	*1950–78 annual growth (percent)*	*1978–95 annual growth (percent)*
Jiangsu	334	1,291	11,129	4.9	13.5
Sunan	454	2,000	25,983	5.4	16.3
Suzhou	430	1,865	31,039	5.4	18.0
Wuxi	454	2,134	25,755	5.7	15.8
Changzhou	498	2,066	17,576	5.2	13.4
Suzhong	338	1,404	10,995	5.2	12.9
Nanjing	285	2,450	15,357	8.0	11.4
Zhenjiang	303	1,380	13,037	5.6	14.1
Yangzhou	325	959	10,100	3.9	14.9
Nantong	393	1,343	8,486	4.5	11.5
Subei	260	852	4,667	4.3	10.5
Huaiyin	236	723	3,446	4.1	9.6
Lianyungang	245	936	5,115	4.9	10.5
Xuzhou	250	932	4,477	4.8	9.7
Yancheng	309	900	6,269	3.9	12.1

Source: JSB 1994b; 1996.
Note: 1990 constant prices.

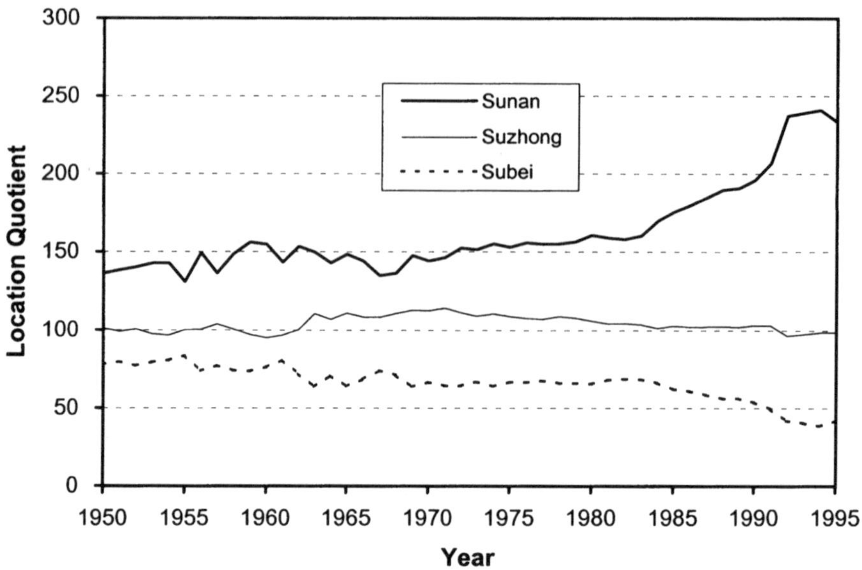

Figure 5.5 Change of location quotients in Sunan, Suzhong, and Subei, 1950–95.

slightly, as the growth of Sunan was matched by agricultural improvement in Subei. However, since the mid-1980s, with the emphasis of reforms shifted to comprehensive reforms and open door policies, Sunan outgrew Subei and Suzhong by a great margin, and the Sunan–Subei divide has been intensified. Figure 5.5 shows the dramatic rise of the relative status of Sunan after the mid-1980s, and the relative decline of Subei's.

After a slowdown of rising interregional inequality in 1989 due to the Tiananmen incident and the related policy change, interregional inequality soared in the early and mid-1990s, as shown in Figs 5.4 and 5.5. Sunan registered much faster economic growth than the rest of Jiangsu: the CV reached 0.64 in 1995, doubling its 1978 value.[12] By the mid-1990s, the north–south divide in Jiangsu was larger than ever during Jiangsu's post-revolution development, although the extent of rising inequality has been stabilized since the mid-1990s.

5.3.2 Intercounty and rural inequalities

Economic development is also uneven among county-level administrative units. As shown in the changes in the coefficient of variation (CV), we see a steady increase of intercounty and rural inequalities during Mao's period as a whole (Fig. 5.4). In 1978, the CVs for intercounty and rural inequalities were 1.13 and 0.45, respectively, which were higher than those in 1950 (0.34 and 0.33, respectively). Two major periods recorded rising inequality: from the late 1950s to early 1960s (especially between 1961 and 1963), and during the 1970s. During the reform period, rural inequality has increased further; intercounty inequality, however, has declined.

During the early and mid-1950s, China emphasized the recovery of traditional industrial bases in the coastal region and the development of the interior region (Ma and Wei 1997). Jiangsu was de-emphasized in China's national development planning and resource allocation. With less resources committed to the rich counties in Sunan, rural inequality did not change much. The disastrous Great Leap Forward (1958–60) brought deep economic recession in Jiangsu in the late 1950s and early 1960s. Intensified by bad weather and flooding in Subei in the early 1960s, the poorest agricultural counties, such as Suqian, Pixian, Binhai, and Guannan, were hit the hardest. Their per capita GVIAO declined dramatically during 1962–63, while rich cities and counties in Sunan, with stronger economic bases, still recorded growth. As a result, intercounty inequality rose sharply in 1962 and 1963.

During the Recovery Period (1963–65), China launched economic recovery programs. With the improvement of irrigation and agricultural production, the conditions of the poorest counties in Subei recovered. The provincial government provided preferential policies especially to the poorest counties, including Binhai, Funing, Lianshui, and Guangnan, which greatly helped their recovery (Editorial Committee 1989). For example, per capita GVIAO in Binhai declined from 218 yuan in 1961 to 150 yuan in 1962 and 152 yuan in 1963, but increased to 236 yuan in 1964. Other poor counties also experienced such turn around. These changes led to the decline of intercounty and rural inequalities in Jiangsu, as shown in Fig. 5.4. During the 1970s, a new round of decentralization and the emergence of a pragmatic leadership stimulated the development of TVEs in Sunan. Subei, however, still focused on agricultural production, which grew much slower than industry. As a result, intercounty and rural inequalities in Jiangsu increased.

During the post-Mao reform period, rural inequality has risen rapidly, as shown in Fig. 5.4. The post-Mao period can be divided into two sub-periods: before and after 1984/85. From 1978 to 1984, regional inequality did not change much, corresponding to rural reform in China (decollectivization) which was relatively egalitarian in nature and helped rural development in Subei. This finding is consistent with other findings that earlier years of economic reforms did not intensify rural intercounty inequality (e.g., Lo 1990; Bramall and Jones 1993).

However, with the emphasis of reforms shifting to urban reforms and open door policies in the mid-1980s, rural inequality rose drastically, as shown in Fig. 5.4. By any standard, the speed of rising rural inequality in Jiangsu is dramatic. After a slight setback related to the 1989 Tiananmen incident and with the deepening of economic reforms, rural inequality has intensified further in Jiangsu. CVs confirm little change from 1978 to 1984, but have increased dramatically subsequently. In 1984, the CV was 0.46, which was only slightly higher than in 1978 (0.45). But in 1995, the CV rose to 1.04, more than twice as high as its level in 1984 and the highest level during the study period (1950–95).

While rural inequality has been intensified during the reform period, intercounty inequality actually declined: the CV for intercounty inequality declined from 1.17 in 1977, the highest value and right before the reform in 1978, to 0.85 in 1992 and 1993, the lowest level during the reform period. The decline of CVs was not the result of the government's intention of balancing regional development; rather, it is due largely to the slower growth of cities and the emergence of counties in Sunan. I will further demonstrate the changing forms of regional development in Jiangsu, and explain factors underlying the slower growth of traditional cities and the faster growth of counties in Sunan.

5.4 New forms of regional development

5.4.1 Where the grass is greener

To further illustrate the trends of regional inequalities analyzed above, I calculated and mapped the growth rates of GVIAO per capita in counties. During Mao's era, the growth rate of GVIAO per capita in Sunan was only slightly higher than in Suzhong and Subei (Table 5.4). From 1978 to 1995, however, Sunan's GVIAO per capita grew annually at 16.3 percent, much higher than that of Subei and Suzhong (10.5 percent and 12.9 percent, respectively). In 1995, Sunan's GVIAO per capita reached 25,983 yuan, while that in Subei was only 4667 yuan. Subei, with 42.8 percent of the population of Jiangsu, produced only 22.6 percent of the province's GDP.

Gaps among counties in Jiangsu were large in the 1950s, shortly after the establishment of the new republic (Fig. 5.6). Sunan in general had higher GVIAO per capita. From 1950 to 1978, as shown in Fig. 5.7, while the major gap in growth was between cities and counties (those with smaller areas are cities), more counties in Sunan grew faster than 4 percent annually than in Subei. In general, rich counties recorded faster growth than poorer counties. In 1978, while the conditions of the poorest counties were improved somewhat, many counties in Subei remained the poorest (Fig. 5.8), as growth in Subei mainly took place in the cities. Overall, although patterns of industrial and agricultural output in 1978 appeared more dispersed than in 1950, gaps between the cities and the countryside and, therefore, between Sunan and Subei increased. As a result, regional inequalities rose during Mao's era. Those patterns of uneven regional growth confirm findings revealed by CVs and LQs.

During the reform period, counties in Sunan recorded much faster growth than Subei (Fig. 5.9). The rapid growth of counties in Sunan has contributed to the intensification of interregional and rural intercounty inequalities. Meanwhile, cities (city districts) recorded slower growth than many counties, leading to the decline of intercounty inequality. Indeed, large gaps exist

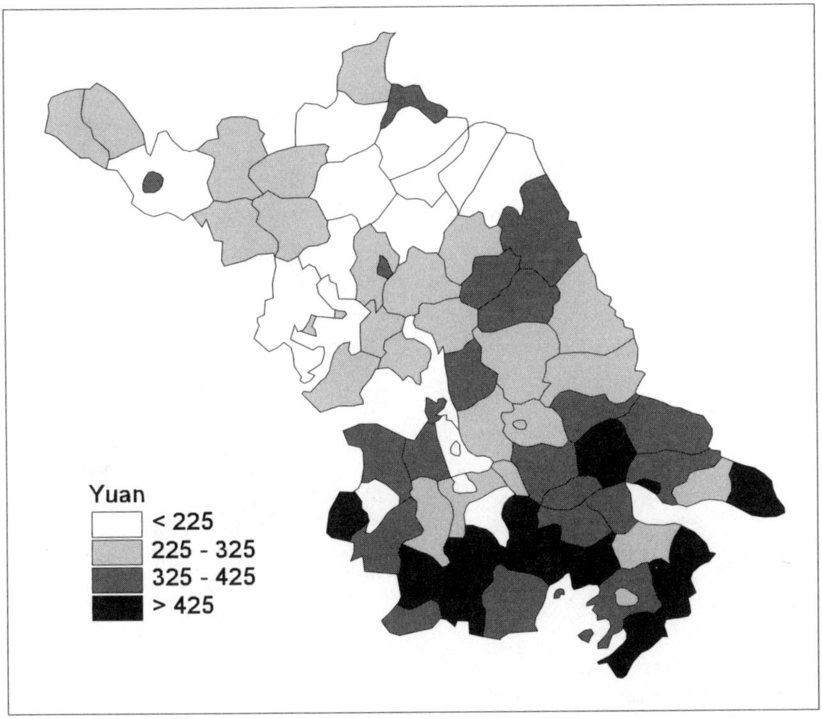

Figure 5.6 GVIAO per capita in Jiangsu, 1950.

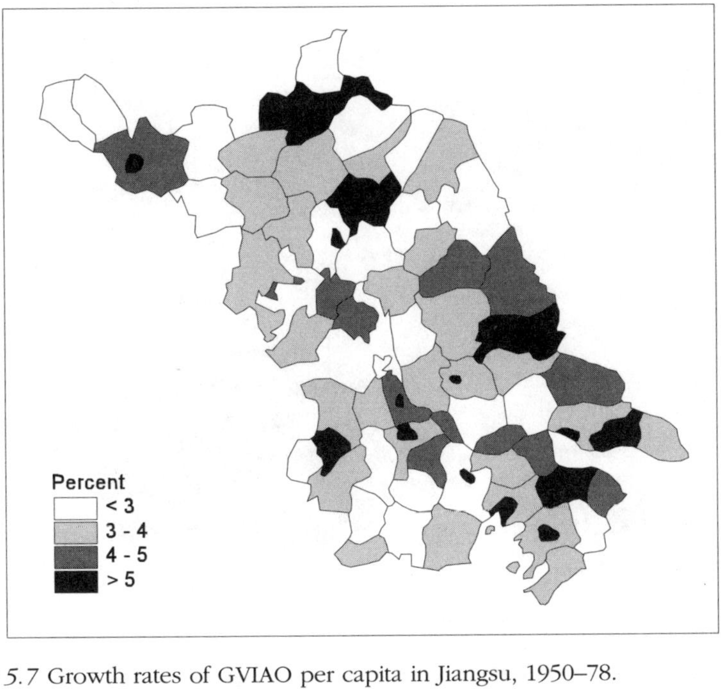

Figure 5.7 Growth rates of GVIAO per capita in Jiangsu, 1950–78.

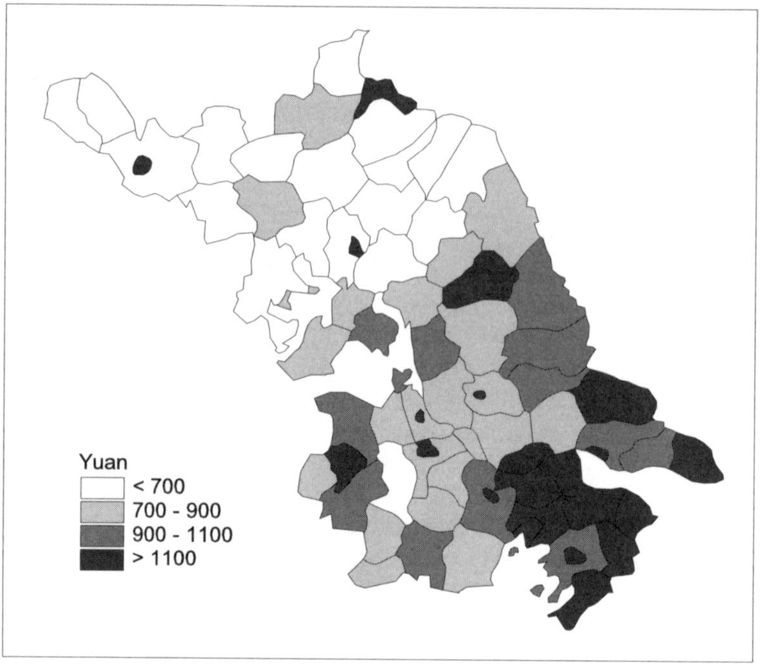

Figure 5.8 GVIAO per capita in Jiangsu, 1978.

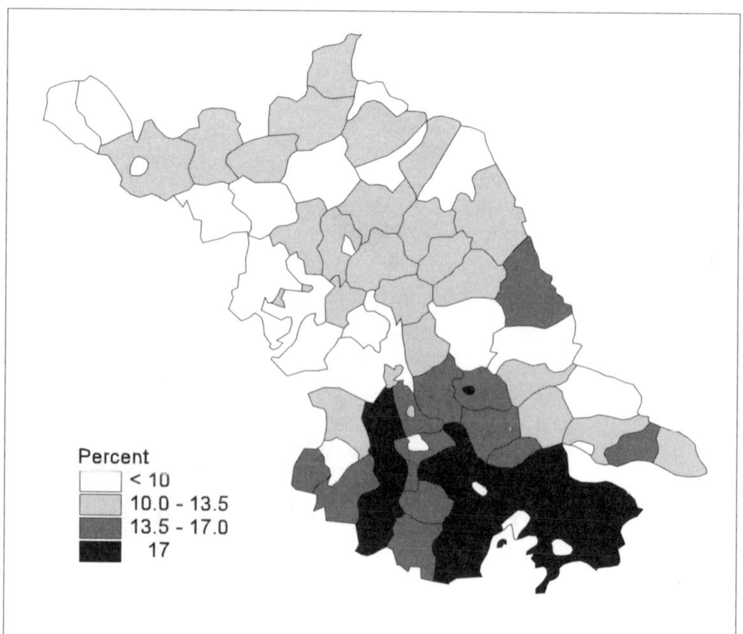

Figure 5.9 Growth rates of GVIAO per capita in Jiangsu, 1978–95.

between the rich and poor regions (Fig. 5.10). In 1995, all of the 12 counties (including county-level cities) in Sunan were among the richest in China. Counties in Subei, such as Binhai and Guanyun, were still among the poorest. Some counties in Sunan even have per capita GVIAO higher than most cities in Subei.

To further illustrate the gap among Sunan, Suzhong and Subei, I have graphed the changing status of municipalities from 1950 to 1995, based on location quotients. As shown in Fig. 5.11, the status of the three municipalities in Sunan has increased greatly since the mid-1980s. The richest counties in Sunan, including Wujin, Jiangyin, and Wuxi (Xishan), have grown rapidly, while the cities of Wuxi, Suzhou, and Changzhou recorded slower growth. Correlation analysis reveals that a dummy variable representing cities is negatively related to growth rates of GVIAO per capita ($r = -0.44^{**}$).[13] By 1995, per capita GVIAO of the richest counties even surpassed the cities, and their per capita FDI was also among the highest.

Counties in Suzhong are in the transition between Sunan and Subei (Fig. 5.12). While the status of Yangzhou and Zhenjiang municipalities improved, that of Nantong and Nanjing declined somewhat, partially related to the problems with SOEs. Within Suzhong, some counties, such as

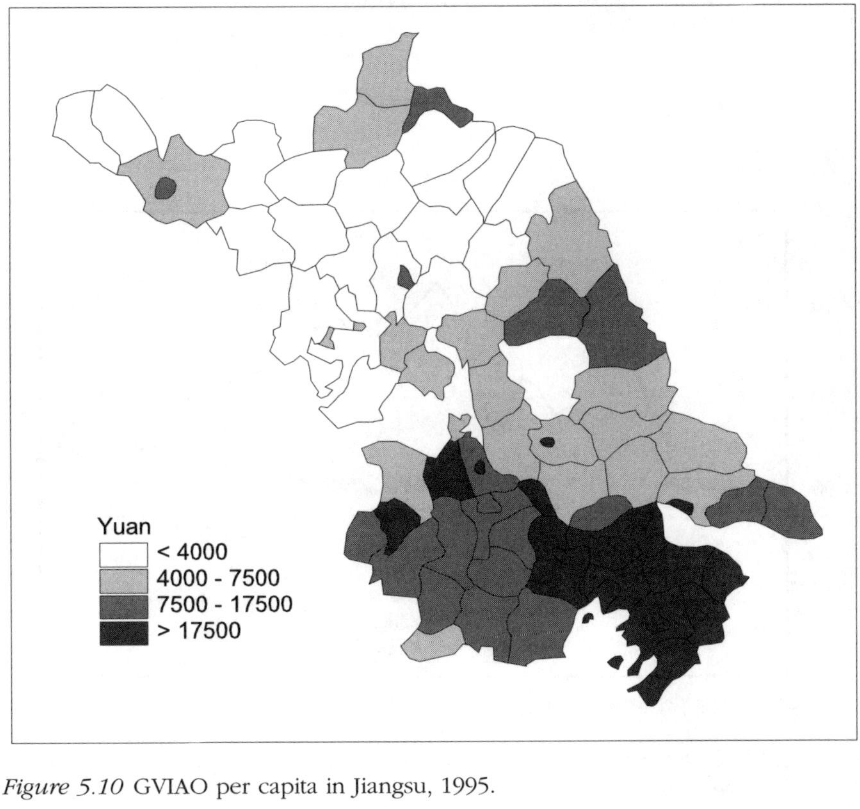

Figure 5.10 GVIAO per capita in Jiangsu, 1995.

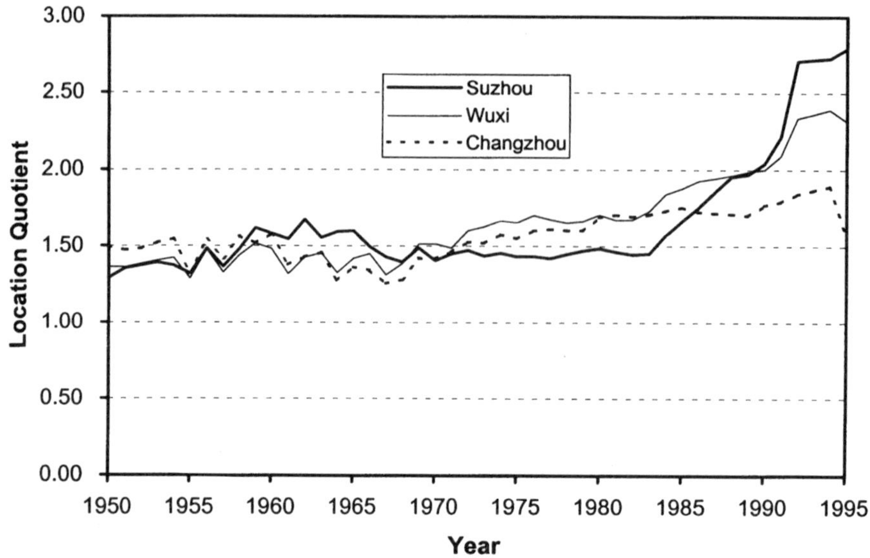

Figure 5.11 Location quotients based on GVIAO per capita for municipalities in Sunan, 1950–95.

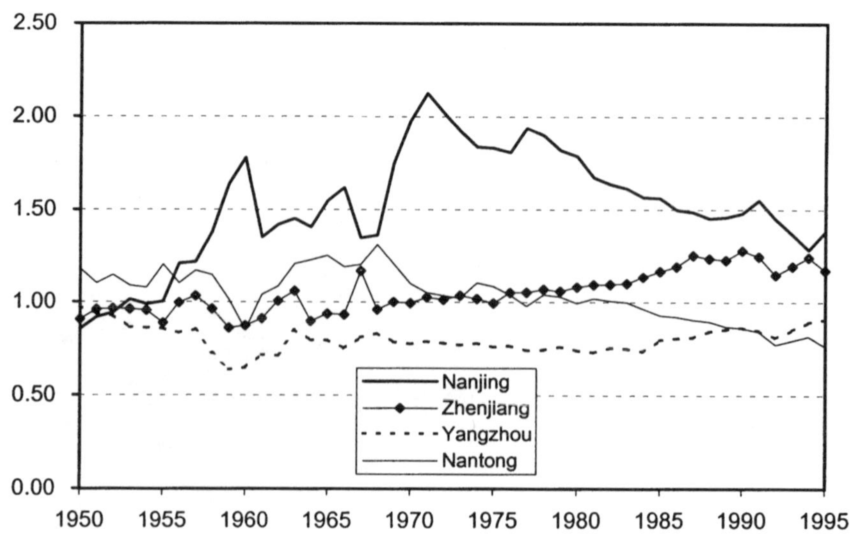

Figure 5.12 Location quotients based on GVIAO per capita for municipalities in Suzhong, 1950–95.

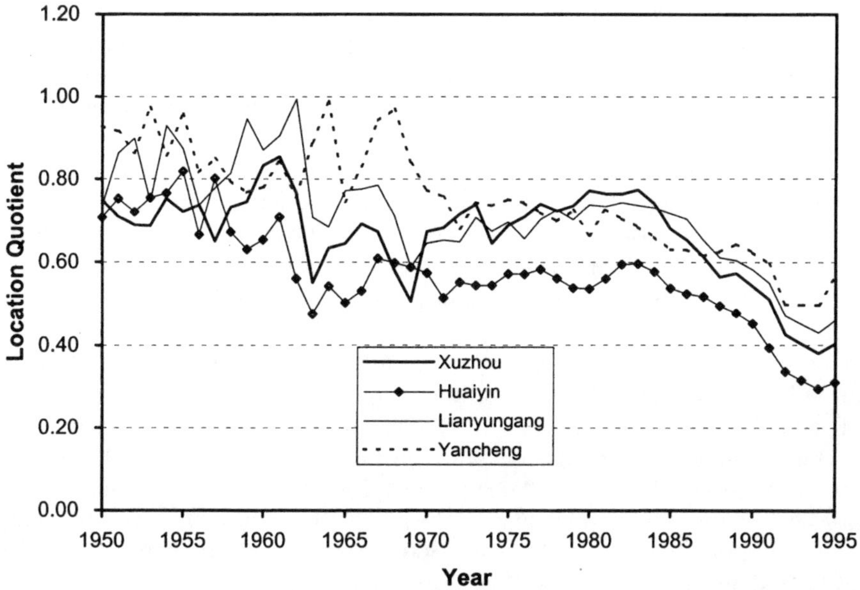

Figure 5.13 Location quotients based on GVIAO per capita for municipalities in Subei, 1950–95.

Jingjiang, are catching up with counties in Sunan with their rapid industrial growth and close ties to Sunan and Shanghai, while the status of some other counties (e.g., Gaoyou) is still falling behind.

All municipalities in Subei have experienced declining status (Fig. 5.13). The status of counties in Subei has fallen, especially since the mid-1980s and during the 1990s. For example, while Fengxian's status in the economy of Jiangsu declined somewhat during Mao's era, its decline since the mid-1980s is substantial. The status of Ganyu and Shuyang also experienced dramatic decline. The rapid growth of Sunan and the slow growth of Subei have intensified interregional and rural inequalities in Jiangsu.

In summary, during Mao's era, regional inequalities in Jiangsu increased slightly, although the conditions of the poorest counties were improved somewhat. During the reform period, particularly after the mid-1980s, the richest counties in Sunan, such as Wujiang, and Kunshan, recorded much faster economic growth, while counties in Subei fell far behind, leading to the intensification of interregional and rural inequalities. Intercounty inequality, however, has declined mainly due to the declining status of old cities.

5.4.2 Development paths of selected regions

I have selected four cities and counties to illustrate the changing dynamics between cities and counties in Jiangsu: Wuxi, Kunshan, Shuyang, and

Gaoyou. The city of Wuxi (Wuxi shiqu or Wuxi city district) is selected to represent cities with higher per capita income. Kunshan is selected to represent counties in Sunan, which have grown rapidly during the reform period. Shuyang is one of the poorest counties in Subei. Gaoyou is selected to represent counties in Suzhong, which have experienced some growth but still fall behind Sunan.

During Mao's era, Wuxi shiqu experienced faster growth than counties in Jiangsu. Wuxi, like many other Chinese cities, benefited from Mao's policy favoring the development of industries and cities. Wuxi's growth largely resulted from state financial support, cheap energy and raw materials, and highly priced industrial products, although Mao in certain periods advocated policies to encourage rural development. As a result, the economic status of Wuxi in the economy of Jiangsu rose greatly, as shown in Fig. 5.14. Wuxi's growth reflected faster growth and the rising status of cities in the economy of Jiangsu.

But the status of Wuxi has declined during the reform period. During the 1980s, Wuxi's GDP grew by 247.9 percent, lower than all of the three counties within the Wuxi municipality (Xishan (Wuxi county), 476.5 percent; Jiangyin, 538.9 percent; and Yixing, 360.6 percent) (Wuxi 1996, p. 348). In 1994, Xishan's GDP reached 17.7 billion yuan, surpassing that of Wuxi (16.8 billion yuan). The slow growth of Wuxi is partially related to strong government control, heavy tax burdens, the low efficiency of SOEs, and an overburdened urban infrastructure. Wuxi's economy has been dominated by less efficient and slower growing SOEs. In 1980, SOEs produced 81 percent of

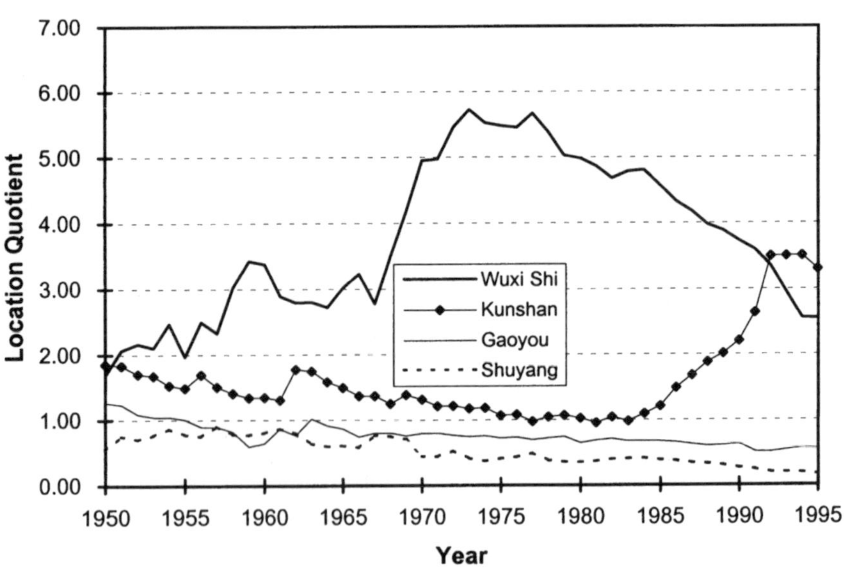

Figure 5.14 Location quotients based on GVIAO per capita in selected counties, 1950–95.

industrial output, and in 1990, SOEs still absorbed 58.4 percent of urban workers and produced 61.8 percent of industrial output, which were considerably higher than the counties (SSB 1994). Its industrial growth has also been handicapped by deteriorating industrial equipment, the lack of energy and raw materials, rising labor cost, and limited space for expansion. However, Wuxi municipality's total tax contribution to the state is at an amount similar to Guangdong province. Meanwhile, the growth of the service and private sectors in Wuxi has been slow. In 1990, the tertiary sector only produced 29.3 percent of GDP. Although the city has attracted considerable foreign investment and enlarged its administrative area, its economic status has been declining during the reform period (Fig. 5.14). Wuxi's share of GDP of the Wuxi municipality declined from 48.3 percent in 1980 to 37.2 percent in 1990 and 28.3 percent in 1995.

The declining status of central cities has intensified the city–county relationship in Jiangsu. Cities have used their administrative power to expand their land areas, which were previously under the administration of the suburban counties. Measures have been taken to increase retention of revenue and to attract more investment from foreign investors. With the establishment of the Wuxi Singapore Industrial Park, FDI in Wuxi increased from US$3.48 million in 1990, which was less than Wuxi county (US$3.5 million), to US$328 million in 1996. Efforts have also been made to develop the tertiary sector and to revitalize the problematic SOEs.

On the other hand, several counties in Sunan have grown dramatically. Correlation analysis shows a close positive relationship between counties located in Sunan and growth rates of GVIAO per capita. During Mao's era, counties in general were less favored than cities in economic development. Although some counties in Sunan recorded faster growth than counties in Subei, their growth rates were still much lower than cities. During the reform period, however, counties in Sunan have experienced dramatic growth—faster than old cities and Subei's counties. The phenomenal growth of TVEs, in particular, enhanced by the infusion of foreign investment, has greatly contributed to the growth of counties in Sunan.

Kunshan, one of the richest counties in Jiangsu, for example, has grown rapidly, especially since the mid-1980s. By 1995, as shown by Table 5.5,

Table 5.5 Major economic indicators in selected areas, 1995

	Population (thousands)	*FAI (million yuan)*	*Per capita FAI (yuan)*	*FDI (US$ million)*	*Per capita FDI (US$)*	*Per capita GVIAO (yuan)*
Wuxi shiqu	1,074.8	9,745	9,067	328.3	305.5	28,435
Kunshan	580.5	4,498	7,748	505.3	870.5	38,982
Gaoyou	831.5	376	452	3.1	3.7	6,287
Shuyang	1,569.3	149	95	3.0	1.9	2,078

Source: JSB 1996.

its per capita GVIAO surpassed the city of Wuxi, and per capita FDI was more than double that of Wuxi. As I will elaborate in the next chapter, several factors have contributed to the rapid growth of Kunshan, including China's reform policies, the importance of local states and development conditions, and the emergence of foreign investment. As a result of the rapid growth of counties in Sunan and the slower growth of cities, intercounty inequality declined during the reform period. The rapid growth of counties in Sunan is not the deliberate attempt of the government to develop counties and to reduce urban–rural inequality; rather, it is a byproduct of economic reforms that stimulated the growth of non-state sectors and counties, and reform policies have also been more favorable to counties.

The status of counties in Subei (represented by Shuyang) has fallen, especially during the 1990s. Shuyang's per capita GDP, per capita GVIAO, and per capita revenue are all among the lowest in Jiangsu. Compared with Sunan, Shuyang's economy is dominated by agriculture and SOEs, which grow slower and are less efficient in production. Meanwhile, the county has to maintain spending in education, health care, infrastructure, and governments. As local spending has always been higher than local revenue, it could not even balance its budget and has to be subsided by the provincial government. The county lacks capital for investment, as per capita fixed investment and FDI in Shuyang were considerably less than Kunshan (Table 5.5). From 1990 to 1996, Shuyang only attracted US$5.68 million FDI, and is among those counties in Jiangsu with the lowest FDI.

As shown in Fig. 5.14, the gap between Kunshan and Shuyang has widened rapidly. The rapid growth of Sunan and the slow growth of Subei have contributed to the intensification of interregional and rural intercounty inequalities in Jiangsu. Per capita income in several counties in Sunan has even surpassed the cities, and the rapid growth of these leading counties has significantly contributed to the decline of intercounty inequality.

Counties in Suzhong (represented by Gaoyou of Yangzhou municipality) caught up with Sunan somewhat in economic growth. Based on scale economies, Yangzhou has developed a series of nationally known manufacturing products, such as air conditioners (Chunliang Group), buses (Yangzhou Bus Factory), trucks, and medicine.[14] Meanwhile, the growth of TVEs has been accelerated. But dramatic gaps exist among counties in Suzhong, especially in Yangzhou (Zhang and Wei 1986). On the one hand, counties in the southern part of Suzhong, such as Jingjiang, have recorded rapid growth, especially through the development of TVEs; their growth patterns very much resemble counties in Sunan. Jingjiang has established close economic and social ties with Sunan and Shanghai, and has benefited greatly from interfirm cooperation with these areas. On the other hand, counties in the north, such as Xinghua and Baoyin, have lagged behind in growth; their governments have had strong desires to promote economic growth, but the local development conditions are less favorable than counties in Sunan. Agricultural decline had led some agricultural counties in Suzhong into fiscal distress (Rozelle 1996).

Gaoyou is situated between those more rapidly growing counties and those slowly growing counties in Suzhong. Gaoyou's economy was dominated by agriculture, with few industrial enterprises. Its economic status in Jiangsu declined during Mao's era. From 1952 to 1978, its average annual growth rate of per capita GVIAO was only 2.8 percent, lower than the provincial average. Gaoyou's economic growth has been accelerated during the reform period, including the growth of TVEs. But Gaoyou is under intense competition from the more developed Sunan and also from Subei.

The different development paths of cities and counties in Jiangsu show that the wealth of regions is changing, although slowly. With the changing domestic and international conditions for development, some of the richest regions (i.e. the cities) have lost the leading status they enjoyed during Mao's era, while some counties have successfully taken opportunities to improve their economies. While interregional, intercounty, and rural inequalities have shown different patterns of change, the core–periphery structure in Jiangsu has persisted over the years.

5.5 Conclusion

This chapter has examined the development process and changing patterns of regional development and inequality in Jiangsu. I have found that patterns of interregional inequality differ from patterns of intercounty and rural inequalities. During Mao's era, contrasting to the normal expectation that spatial inequality declined in China, I have found that interregional, intercounty, and rural inequalities increased in Jiangsu, due largely to the rapid growth of cities and Sunan. The increase of intercounty inequality was especially rapid. This finding is consistent with the finding at the national scale, i.e. both interregional and interprovincial inequalities increased or at least persisted in China.

During the reform period, interregional and rural inequalities have increased further, and the gap between Sunan and Subei has been intensified, as richer counties in Sunan registered more rapid growth than poorer counties in Subei. Counties with more SOEs and heavy industry, largely in Subei, have grown more slowly than counties in Sunan, which are based on the development of non-state enterprises. Intercounty inequality, however, declined, due mainly to the slower growth of old cities than counties. The change of regional inequality in Jiangsu cannot be simply described as convergence, divergence, or inverted-U patterns. In the next chapter, I explain in detail the forces underlying changing patterns of regional development and inequality in Jiangsu.

6 Mechanisms of regional development in Jiangsu

In the last chapter, I have shown the changing patterns of regional development in Jiangsu. In this chapter, I analyze and synthesize the factors underlying these changes. As I have pointed out in the last chapter, previous analyses tend to attribute the development of Jiangsu, especially Sunan, to the success of township and village enterprises (TVEs). In this book, I have argued for the analysis of the central state, local states and agents, and global forces for a better understanding of regional development and inequality in China. While TVEs have indeed played important roles in development of Sunan, I argue that multiple mechanisms, including state policy, local initiatives and factors, and global forces, are important forces underlying the growth of TVEs and regional development in Jiangsu. My analysis in this chapter emphasizes fiscal systems, investment distribution, ownership restructuring, and foreign direct investment (FDI).

6.1 The state, fiscal systems, and regions

6.1.1 The state, local states and agents, and regional development

State policy has played an important role in China's regional development. During Mao's era, development policy favored selected coastal provinces that had strong industrial bases such as Liaoning and Shanghai, but generally avoided other coastal provinces, and redistributed some resources to the interior for the consideration of national defense, industrialization, and equality. Jiangsu, a non-municipality coastal province lacking natural resources, received little investment from the central government; however, like many other coastal provinces, Jiangsu still had to remit huge revenues to the state.

Influenced by the Soviet model of industrialization, Mao's industrial policy favored urban-centered large-scale industrial development, and as a result much investment went to SOEs in the cities. Cities enjoyed faster economic growth than rural areas (often represented by counties). As Jiangsu's enterprises were largely small in scale and a large proportion of them were collectively owned, Jiangsu benefited less from Mao's industrialization policy than coastal provinces such as Liaoning and Shanghai. With little investment from the central government, fewer SOEs were established in Jiangsu.

However, during certain periods the state encouraged economic decentralization and rural development, which stimulated the growth of rural enterprises in Jiangsu, particularly in Sunan (Ho 1994). As small enterprises dominated Jiangsu, its economy was less controlled by the central and provincial governments and survived better than many other provinces during the Cultural Revolution. Moreover, the provincial government of Jiangsu decentralized most of enterprise management functions to lower-level governments that had great incentives to develop TVEs, as TVEs created jobs and more profits could be retained locally.

Post-Mao economic reforms have stimulated the growth of Jiangsu and benefited Sunan in particular. New fiscal policies allowed richer areas in Sunan to retain some of their increased revenues. Economic reforms have also given regions greater autonomy in investment decision making, especially in terms of financing investment and investment allocation strategies. With declining state control and increasing incentives for growth, local governments and agents can mobilize more capital for economic development. Some regions, especially those experiencing more rapid growth, are able to channel more funds for investment mainly to more profitable sectors. Moreover, post-Mao economic reforms have provided favorable conditions for the development of non-state enterprises, especially TVEs. The open door policy has further boosted the growth of Jiangsu's economy, as Jiangsu was among the provinces that were opened up earlier for foreign investment and trade.

While the central state remains important, local states and agents have become important forces driving economic development, as a result of economic reforms and globalization. First, local states have played a significant role in promoting regional development in Jiangsu. The achievement of township and village officials is largely dependent on the success in economic growth, particularly in the development of TVEs (Oi 1992; Walder 1995b). Lower-level governments in Jiangsu have actively promoted the development of local economies (Ma and Fan 1994). They have bargained intensely with higher-level governments for more decentralized policies, and have initiated new policies to attract foreign investment and to promote international trade. Moreover, TVEs have been considerably less controlled by planning and rely heavily on markets, and profits from these enterprises are also largely retained by localities. Local cadres have become bureaucrats–entrepreneurs who actively seek and accumulate resources, such as capital, raw materials, technology, and markets that are necessary for local economic development.[1]

The provincial government of Jiangsu has actively bargained with the central government to obtain more favorable reform policies. It has also been directly involved in negotiating with foreign investors and has offered generous investment incentives. Moreover, the provincial government has implemented a decentralized system that gives considerable decision-making power to the governments of cities, counties, and townships. It has facilitated economic reforms by promoting successful reform experiences. For example, Yanqiao township of Wuxi county (in Sunan) created a system of '*yibao sangai*' (one contract and three reforms) which introduced market mechanisms into the management of TVEs.[2] Their experiences in initiating local policies were quickly promoted by the provincial leaders and implemented in many TVEs throughout the province (Gu and Shen 1989).

Secondly, the development of Jiangsu, Sunan in particular, has benefited from the economies of nearby cities, especially Shanghai. The development of rural industries is closely related to cities that provide investment, markets, subcontracting linkages, and technical know-how.[3] Urban and rural areas in Sunan have traditionally had close economic linkages with each other, and numerous small towns in Sunan linked the rural economy to the cities in the region (Skinner 1977; Johnson 1993). These historical ties between cities and the countryside have facilitated the development of Sunan, mainly through the growth of TVEs. The economic ties to Shanghai have been particularly important to the development of Sunan. Besides Shanghai, TVEs have established close ties with many other cities, particularly Suzhou, Wuxi, and Changzhou.

Thirdly, development paths of the regions in Jiangsu have important implications for their growth patterns. During Mao's era, central, provincial, and local governments stressed agricultural production in Subei, and few resources were committed to the development of light industry and TVEs. These development experiences partly contributed to their diverse development paths and achievements during the reform period. While Sunan has grown more rapidly through industrialization and the development of non-state sectors, Subei has lagged behind due to the dominance of slowly growing agriculture and SOEs.

Lastly, local resource endowments are important to the development of TVEs, which have been less supported by the central government and have to rely more heavily on local resources and markets than SOEs. Due to the lack of state investment, TVEs also rely on local resources to finance construction and production. Sunan has long been one of the most productive agricultural bases in China, providing capital and raw materials for the development of TVEs. These local advantages help TVEs develop strong abilities to adjust to the constantly changing macro economic policies and markets in China. These factors of regional development in Jiangsu will be elaborated in the following sections.

6.1.2 Fiscal systems, budget, and regions

After the establishment of the PRC in 1949, China in the early 1950s established centralized fiscal and investment systems to take control of the Chinese economy and to finance industrialization. Like many former socialist

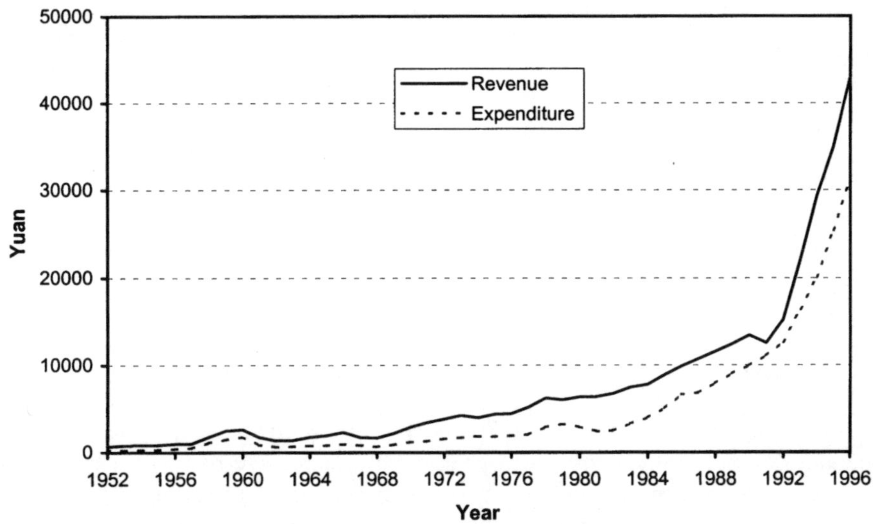

Figure 6.1 Revenue and expenditure in Jiangsu, 1952–96.
Source: JSB 1991–97.

countries in Eastern Europe, the fiscal system under Mao was centralized (Wei 1996), although central–local fiscal arrangements changed over time. The emphasis of resource allocation was in the traditional industrial bases and the interior provinces. Jiangsu, like many other coastal provinces, submitted huge revenues to the central government, but received much less 'downward' fiscal transfers.

During Mao's period, Jiangsu's expenditure was always significantly less than its revenue, due to the large transfers to the central government (Fig. 6.1). During the recovery period, Jiangsu handed over 76 percent of its revenue to the central government (Table 6.1). In 1952, Jiangsu's revenue was 680 million yuan, considerably larger than expenditure (270 million yuan). In 1955, local revenue in Jiangsu reached 847 million yuan, while local expenditure was only 289 million yuan. In 1965 when China's Third Front project was initiated, Jiangsu's expenditure was 8 billion yuan, again much less than revenue (19.4 billion yuan). In 1976, on the eve of economic reforms, Jiangsu's revenue reached 4.4 billion yuan, but its expenditure was only 2.49 billion yuan. Although Jiangsu received some decision-making power during the late 1950s and early 1970s, which stimulated local efforts in revenue generation and increased local spending, the gap between revenue and expenditure remained large. From 1952 to 1976, the province's expenditure was 24.11 billion yuan, which only accounted for 43.9 percent of its revenue (54.92 billion yuan). Jiangsu's per capita expenditure during Mao's era was among the lowest in China.

Table 6.1 Jiangsu's revenue contribution to the state

Period	*Amount (billion yuan)*	*As percent of revenue*
Period of Recovery (1949–52)	1.22	76.0
First Five-Year Plan (1953–57)	2.75	61.6
Second Five-Year Plan (1958–62)	4.20	42.2
Period of Readjustment (1963–65)	2.83	56.1
Third Five-Year Plan (1966–70)	6.35	59.0
Fourth Five-Year Plan (1971–75)	11.77	59.7
Fifth Five-Year Plan (1976–80)	16.66	59.8
Sixth Five-Year Plan (1981–85)	23.48	64.2
1995	19.00	54.2

Source: Editorial Committee 1989; JSB 1996.

After the death of Mao in 1976, new experimental fiscal systems were initiated to provide incentives to regions and to reduce fiscal burdens. The provincial leaders proposed to the central government a reform package that would give greater leeway to Jiangsu in economic planning, resource allocation, and fiscal systems. Consequently, in 1977 Jiangsu implemented a new fiscal system of 'fixed rate contracts' (*guding bili baogan*), under which the rate of revenue sharing was fixed for four years, and the province also received more authority in budgeting and expenditure. This system clearly stimulated revenue generation and spending, and in 1977 and 1978, both revenue and expenditure grew rapidly in Jiangsu. But Jiangsu's revenue transfer to the central government remained large, as in 1978, Jiangsu's per capita revenue was 105 yuan, ranked 7th in the country, but its expenditure was only 49 yuan, ranked the 22nd.[4]

Based on the experiments of Jiangsu, Sichuan, and other provinces, the central government decided in 1979 to launch national fiscal reforms in 1980. The 1980 fiscal system provided the provinces more fiscal powers in revenue sharing and expenditure, following the idea of 'eating in a separate kitchen.' Jiangsu continued its four-year fixed revenue sharing with some adjustment, which was slightly changed as 'sharing total revenue' (*zonge fencheng*) in 1981. But Jiangsu's local revenue retention rate was still among the lowest in China (39 percent), only higher than a few other coastal provinces (e.g., Shanghai, Zhejiang, and Shandong). Its local retention rate was even further reduced during 1983 and 1984 (about 35.7 percent), as the central government demanded a reduction in local expenditure and borrowed from localities (Shen 1993). From 1977 to 1985, Jiangsu submitted 71 billion yuan revenue to the central government (Editorial Committee 1989). The fiscal burden for the province remained high, especially in cities such as Wuxi.

To provide more local incentives, Jiangsu implemented a slightly revised system during the 1985–87 period—'sharing total revenue with growth adjustment' (*zongge fencheng jia zengzhang fencheng*). Another fiscal reform was implemented in China in 1988 to further stimulate local efforts in revenue generation. Jiangsu's fiscal system was improved by introducing 'basic sharing with growth adjustment' (*shouru dizeng baogan*), in which a specified nominal proportion was contracted to the central government with a certain growth rate. While this arrangement initially improved revenue, revenue growth slowed down in the following years. Although reforms in the 1980s provided more incentives and fiscal powers to the provinces, it was clear in the early 1990s, that these reforms did not solve the fundamental problems associated with state socialism, such as intense intergovernmental bargaining, vague budget categories, reliance on administrative control, and budget deficit and debts (Wong *et al.* 1995; Wei 1996).

To protect local interests and satisfy the expenditure need, Jiangsu, like many other coastal provinces, expanded its extrabudgetary funds. The ratio of extrabudgetary revenue to budgetary revenue increased from 24 percent in 1978 to 68 percent in 1985. In 1990, Jiangsu's extrabudgetary revenue was 11.5 billion yuan, only slightly lower than its budgetary revenue (13.6 billion yuan). Its extrabudgetary expenditure was 11.4 billion yuan, which surpassed its budgetary expenditure (10.1 billion yuan). Extrabudgetary funds are largely controlled by localities, benefiting local economies greatly. However, the provincial-level revenue was −110 million yuan, while the provincial-level expenditure was 2.2 billion yuan, indicating that the provincial fiscal capacity was extremely weak (Jiang 1991).

Tensions and problems arising from China's fiscal systems prompted the Chinese leaders to push for fiscal reforms at the Third Plenum of the Fourteenth Central Committee held in November 1993. A new fiscal system was fully implemented in 1994 after the experiments of 1992–93. It is a tax assignment system (*fen shui zhi*) which divides taxes into central, local, and shared ones collected by central or local fiscal agencies. It has more clearly specified fiscal responsibilities of central and provincial governments and partly replaced the traditional bargaining in determining central–provincial revenue-sharing rates. Meanwhile, the central government increased its share of budget revenue and attempted to strengthen its control over fiscal resources.

In 1995, Jiangsu's aggregated revenue was 35 billion yuan (local revenue was 17.3 billion yuan), which was considerably higher than expenditure (25.3 billion yuan). Revenue was mainly from industrial and commercial taxes and financed fixed investment and education. However, administrative spending accounted for 16.1 percent of local expenditure, and 1.2 billion yuan was used to subsidize loss-making SOEs. Extrabudgetary revenue reached 24.9 billion yuan in 1996, more than any other province in China, although Jiangsu's extrabudgetary expenditure was slightly smaller than Guangdong.[5] Jiangsu submitted to the central government 17 billion yuan in value-added tax and 2 billion yuan in consumption tax (People's Government of Jiangsu Province 1996), indicating that the new fiscal system further increases Jiangsu's fiscal contribution to the central government. Jiangsu remains a heavily up-transfer (*shang jia*) province. The heavy revenue burden has weakened the provincial budget, and considerable disparities in fiscal structures exist among regions in Jiangsu.

6.1.3 Regional patterns of revenue and expenditure

Corresponding to the restructuring of the central–provincial system, provincial–local fiscal systems have also experienced a series of changes to provide more incentives to localities in revenue growth. In 1977, Jiangsu led the nation by initiating a system of fixed rate contracts with growth adjustment with the lower governments. During the 1982–87 period, Jiangsu implemented a system of total rate contracts (*quane bili baogang*) with local level governments (Shen 1993). In 1983, townships were allowed to establish their own fiscal systems, and therefore gained more fiscal authorities. By 1985, 85.3 percent of all townships in Jiangsu had already established their fiscal systems (Zheng 1995).

In 1988, the provincial–local fiscal system was changed to basic sharing with growth adjustment (*shouru dizeng baogang*). Further fiscal reforms allowed lower-level governments to retain portions of their increasing revenues and to take more charge of expenditure, stimulating local efforts for revenue generation and spending growth. In addition, local governments also received more powers in tax exemption, especially for foreign investment. But the fiscal burden of cities was considerably higher than counties, although the provincial government attempted to raise local retention rates for cities, such as changing fiscal arrangements with Wuxi, Suzhou, and Nantong in 1985. In 1995, the tax assignment system was implemented in all cities and counties in Jiangsu.

Jiangsu was marked by dramatic regional disparities in revenue and expenditures. Economic reforms have provided considerable incentives for rich counties, although poor counties also received enormous fiscal powers. In general, regions in Sunan generated much more in revenues than those in Subei. Since economic reforms, the gaps between Sunan and Subei in revenue and expenditure have increased, although all regions of Jiangsu have recorded revenue growth. As shown in Table 6.2, Sunan's per capita revenue and expenditure were much higher than that of Subei and Suzhong. Average per capita revenue was 1152 yuan in Sunan, much higher than in Suzhong (670 yuan) and in Subei (271 yuan). The gap in expenditure between Sunan and the rest of Jiangsu was also large (four times larger than Subei). However, Sunan tends to submit more budget surplus to higher-level governments, although it still has higher per capita expenditure.

Several counties in Sunan have had revenues that are among the highest in the nation. In 1980, Changshu and Jiangyin generated revenues of more than 100 million yuan. Wuxi, Wujing, Wujiang, Yixing, and Shazhou soon joined the list. The counties with the highest revenues are all located in Sunan, where TVEs have recorded phenomenal growth. In 1996, 10 counties (including county-level cities) in China had revenues over 400 million yuan,

Table 6.2 Regional disparities in revenue and expenditure, 1996 (billion yuan)

Region	*Revenue*	*Local revenue*	*Expenditure*	*Balance*	*Per capita revenue (yuan)*	*Per capita expenditure (yuan)*
Jiangsu	42.8	22.32	31.1	11.7	602	314
Sunan	15.46	7.47	7.1	8.36	1152	557
Suzhou	6.41	3.25	3.15	3.26	1117	566
Wuxi	5.63	2.56	2.4	3.23	1306	594
Changzhou	3.43	1.66	1.54	1.89	1016	492
Suzhong	16.86	8.19	10	6.86	670	326
Nanjing	7.88	3.62	4.02	3.86	1501	690
Zhejiang	1.78	0.87	1.07	0.71	672	328
Yangzhou	2.04	1	1.43	0.61	458	225
Taizhou	2.08	1.1	1.46	0.62	419	222
Nantong	3.08	1.6	2.02	1.06	392	204
Subei	8.26	4.47	7.08	1.18	271	147
Xuzhou	3.02	1.53	2.14	0.88	352	178
Huaiyin	1.29	0.6	1.08	0.21	263	122
Yancheng	1.89	1.15	1.86	0.03	240	146
Lianyungang	1.3	0.85	1.2	0.1	300	196
Shuoqian	0.77	0.34	0.81	−0.04	160	71

Source: JSB 1997.
Note: Taizhou and Shuqian became separated municipalities in 1996.

five of which were in Jiangsu, including Changshu, Zhangjiagang, Wujin, Xishan (Wuxi), and Jiangyin, all located in Sunan (MOF 1997).

On the other hand, some counties in Subei could not balance their budgets, and required provincial support to reduce fiscal distress. Indeed, in several counties, such as Guannan, Xuyi, Hongze, Shuyang, and Lianshui, during much of the last five decades, revenue could not be balanced. Some counties in Subei have suffered tremendous fiscal difficulties. From 1950 to 1996, Shuyang only balanced its budget in four years: 1952, 1957, 1969, and 1978. The problem of budget deficit was intensified in the 1990s. In 1996, Shuyang's total revenue (including the central revenue) was 120 million yuan, which was considerably lower than its expenditure (190 million yuan), leading to a deficit of 70 million yuan. It is clear that gaps in fiscal conditions are large among counties in Jiangsu.

Meanwhile, cities have also had more difficult fiscal conditions than rich counties, most of which are located in Sunan. While many provinces benefited from fiscal decentralization during the 1980s, fiscal conditions in the cities were less improved. Scholars have reported the fiscal distress and declining economic status of old cities, such as Shanghai, during the 1980s (e.g., White 1989; Wei 1996). Jiangsu's higher revenue submission squeezed the cities that paid higher taxes imposed on SOEs, had lower local revenue retention rates, and were more tightly controlled by higher-level governments in terms of expenditure. Wuxi's revenue submission, for example, was similar in size to the province of Guangdong, which is one of the richest provinces in China and produces more GDP than Wuxi.

6.2 Investment structure and regional patterns

The significance of investment in economic and regional development in former socialist countries is well documented in the literature (e.g., Kornai 1992; Ma and Wei 1997). Investment systems and investment patterns in Jiangsu have changed greatly over time. During the 1950s, China's investment policy favored the recovery of traditional industrial bases and the development of the interior, and most of the coastal provinces, including Jiangsu, were de-emphasized in resource allocation. Among the 156 key-point projects imported from the Soviet Union, none was located in Jiangsu. However, Jiangsu still had to contribute substantial revenues to the state. Moreover, some industrial equipment and technicians in Jiangsu were transferred to poorer provinces for the purposes of interior development and defense. During the 1960s, national defense was emphasized, and heavy industry received high priority in state investment. As the Chinese government favored the interior in allocating defense industries (mainly the Third Front area), state investment in Jiangsu was very limited. During the early 1970s, with the normalization of Sino–US relations and the emergence of a pragmatic leadership, the emphasis of China's regional development began to shift to the coastal region. Several large-scale investment projects were allocated to Jiangsu, particularly in Nanjing. Meanwhile, during the early 1970s, the central government encouraged the development of TVEs, and hundreds of TVEs were established in Sunan.

Since economic reforms, along with other coastal provinces, Jiangsu has received more decision-making power and favorable policies from the central government. Fiscal and investment reforms have provided great incentives for the growth of local economies, and administrative reforms have decentralized many SOEs for local management. Several major changes in fixed investment can be summarized, which very much resemble the changes taking place in China, as I have analyzed in Chapter 3.

First, local governments and enterprises have been given greater autonomy in investment decision making. The reform policies as a whole have defined a climate that encourages active involvement of local officials in economic development and promotes entrepreneurship. The provincial government of Jiangsu has actively bargained with the central government to obtain more favorable reform policies. Moreover, local governments have also obtained considerable power in investment management, and act like local development states in accumulating and mobilizing capital for investment. They entice foreign investors by offering investment incentives more generous than anywhere else in China outside the special economic zones (SEZs) (Pomfret 1995). Capital that previously belonged to commune or

Table 6.3 Sources of finance for fixed investment in Jiangsu (percent), 1985–96

Year	*State budget*	*Domestic credits*	*Foreign investment*	*Enterprise funds*	*Others*
1985	9.0	19.2	2.1	33.6	36.2
1990	4.2	12.8	4.5	29.9	48.7
1993	1.6	22.2	7.8	56.8	11.6
1996	1.2	14.5	17.2	51.4	15.7

brigade-run collectives has become a ready resource for local cadres to mobilize (Fang 1992; Weitzman and Xu 1994). The efforts of local governments are important to the growth and structural change of fixed investment in Jiangsu.

Secondly, regions have obtained a greater access to different sources of investments, such as bank credits, private investment, and foreign capital. Jiangsu has traditionally received lower proportions of investment from state budget, which has been declining rapidly during the reform period. As shown in Table 6.3, from 1985 to 1996, fixed investment financed through the state budget declined from 9.0 percent to 1.2 percent, while investment from enterprise funds and foreign capital increased from 33.6 percent and 2.1 percent to 51.4 percent and 17.2 percent, respectively. This represents a fundamental change in financing regional development in Jiangsu, where proportions of fixed investment coming from the state budget and state-controlled banks have been declining. In other words, non-state sectors and local agents have emerged as major agents of investment, and the central state is playing a less significant role in capital accumulation and investment allocation. The provincial government is also relatively weak in adjusting investment allocation across sectors and regions.

Thirdly, Jiangsu has recorded dramatic growth in fixed investment. During the 1978–95 period, fixed investment grew annually at 29.1 percent. From 1985 to 1995, fixed investment increased from 19.2 billion yuan to 168 billion yuan, and per capita fixed investment increased from 309 yuan to 2378 yuan. The growth rate of fixed investment in the rural areas of Jiangsu was even higher, fueled by the dramatic growth of TVEs. While the rapid growth of investment is related to the high saving rate of Chinese society and the rapid growth of Jiangsu's economy, it is also related to two additional factors:

1 as investment and output are important measures of government achievement, local governments have emphasized increasing investment to stimulate economic growth;
2 many SOEs and some COEs are less concerned with investment efficiency and competition, as efficiency is often a less important criteria for enterprise development (Ni and Wu 1988).

As a result, many regions invested in similar industrial projects, giving little consideration to resources, technology, and competitiveness. While rapid

Table 6.4 Investment allocation in Jiangsu by ownership (percent), 1985–95

	1985	*1990*	*1995*
SOEs	41.9	37.9	35.9
COEs	30.8	21.0	29.2
Others	27.3	41.1	34.9

Source: JSB 1996.

investment growth provides capital for economic growth, it also leads to inflation, the waste of resources, and low investment efficiency.

Lastly, more profitable enterprises, with higher investment returns, have attracted a considerable amount of investment. During Mao's era, much of the investment went to SOEs, mainly for capital construction investment. During the reform period, however, the share of investment in SOEs has declined, due partly to the changing sources of investment and partly to poor investment efficiency in SOEs. The share of fixed investment in SOEs declined from 41.9 percent in 1985 to 35.9 percent in 1995 (Table 6.4). Investment in SOEs has been allocated mainly to projects in energy, transportation, and communications, which take a longer time to build, lack sufficient funds, are poorly managed, and have a poorer production efficiency. On the other hand, local states and agents have been motivated to allocate their investments to more profitable non-state enterprises, especially TVEs and FIEs, which fueled the growth of the non-state sectors. In Jiangsu, TVEs have attracted significant proportions of capital and are an important reason for the province's rapid industrial growth (Prime 1992). In addition, a considerable amount of fixed investment has also been allocated to FIEs, which are among the fastest-growing enterprises in Jiangsu.

While Jiangsu has recorded rapid investment growth, substantial disparities exist within its regions. More developed regions such as Sunan enjoy early implementation of reform policies and are allowed to retain larger shares of their revenues for reinvestment. As shown in Fig. 6.2, the share of Sunan in fixed asset investment has increased during the reform period, while that of Subei has declined. Considerable gaps exist in terms of fixed investment among counties in Jiangsu, as counties in Sunan in general have higher per capita investment (Fig. 6.3). In 1995, fixed asset investment per capita in rural collective enterprises was 1770 yuan in Sunan, and only 315 yuan in Subei. In addition, although cities have higher per capita investment, their investment growth rates tend to be slower, and SOEs and infrastructure took much of the fixed investment.

Sunan and Subei also differ in the allocation of fixed investment in enterprises with different forms of ownership. Since Sunan has a larger non-state sector (79.2 percent compared with 58.1 percent in Suzhong and 59.7 percent in Subei in 1995), it is able to quickly capitalize on the reform policies.

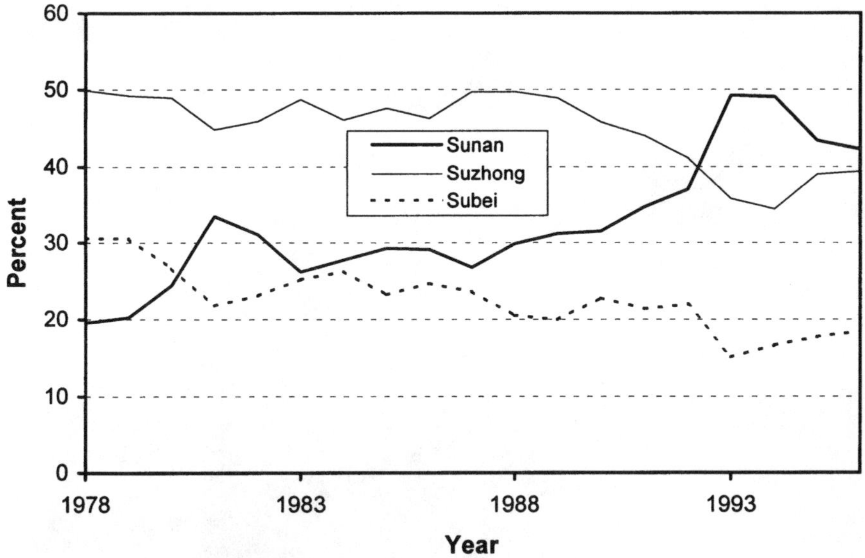

Figure 6.2 Changing shares of fixed investment among regions, 1978–95.
Source: JSB 1994–96; 1994b.

In 1995, in Sunan, only 20.8 percent of fixed investment went to SOEs, while in Subei and Suzhong, more than 40 percent of fixed investment went to SOEs (Table 6.5). Huaiyin of Subei, the slowest-growing municipality in Jiangsu, allocated 55.6 percent of fixed investment to SOEs. In 1995, the ratio of fixed asset investment per capita in SOEs to that in rural collective enterprises was 1.11 in Subei, but only 0.56 in Sunan. As SOEs continue to operate under the pre-reform socialist plans and as their more rigid model of production lags in efficiency and growth, Subei, which has invested more heavily in SOEs, has not shared Sunan's investment efficiency.

As the Chinese economy still relies very much on investment for growth, the regional gap in investment has greatly contributed to regional inequality in Jiangsu. Sunan not only enjoys higher total and per capita investment, its investment also tends to be allocated in more efficient and fast-growing TVEs and FIEs, while poorer Subei invests more heavily in slowly growing and loss-making SOEs. Such regional differences in investment have further differentiated Sunan and Subei in economic structure and in their ability to compete in the market economy.

6.3 Industrialization, ownership, and regions

Although Mao emphasized urban-centered industrialization, during certain periods the state encouraged economic decentralization and rural industrialization, stimulating the development of rural enterprises in Jiangsu,

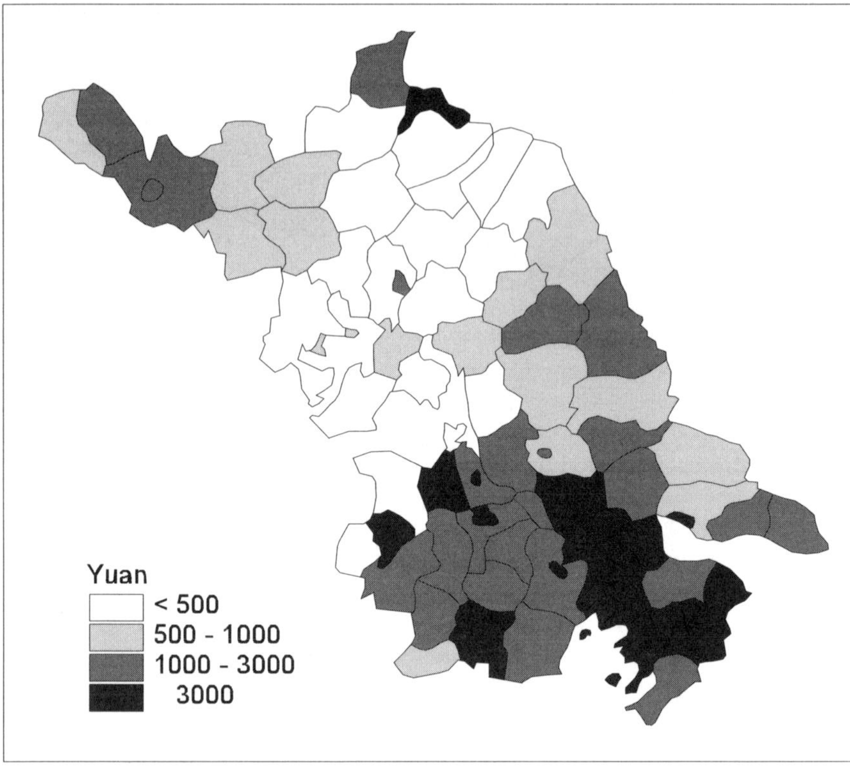

Figure 6.3 Per capita fixed asset investment in Jiangsu, 1995.
Source: JSB 1996.

particularly in Sunan. In 1958, Mao called for rapid rural industrialization. Following his directive, the provincial leaders of Jiangsu required every commune to open 15–25 small enterprises within three to five years in order to achieve this rapid rural industrialization. Consequently, 39,000 commune enterprises were established in Jiangsu (Editorial Committee 1989).

In 1970, during the conference on agriculture in north China, Premier Zhou Enlai called for rapid development of agricultural production. Regions were encouraged to establish new industrial enterprises mainly to support agricultural mechanization and to process agricultural products. These government policies provided the macro policy context for the development of TVEs in Jiangsu and laid the foundation for the phenomenal growth of TVEs in Sunan during the reform period. In 1975, Jiangsu's rural industrial output reached 2.33 billion yuan, more than twice that in 1970. As shown in Fig. 6.4, shares of industrial output produced by SOEs declined rapidly after the early 1970s.

Table 6.5 Regional disparities in fixed investment, 1995 (billion yuan)

		State-owned		*Rural collective*		*Other*	
Region	*Total*	*Volume*	*Percent*	*Volume*	*Percent*	*Volume*	*Percent*
Jiangsu	168.0	60.3	35.9	45.2	26.9	62.6	37.3
Sunan	63.8	13.2	20.8	23.6	37.0	27.0	42.3
Suzhou	30.5	6.0	19.6	10.6	34.8	13.9	45.7
Wuxi	22.7	3.6	16.0	9.0	39.7	10.0	44.2
Changzhou	10.6	3.6	34.4	4.0	37.4	3.0	28.2
Suzhong	57.3	24.0	41.9	12.5	21.7	20.9	36.4
Nanjing	22.4	10.5	47.0	2.7	12.2	9.2	40.8
Zhenjiang	6.1	2.5	41.4	2.2	36.4	1.4	22.2
Yangzhou	15.1	7.2	47.5	3.3	21.8	4.6	30.7
Nantong	13.7	3.8	27.5	4.2	30.7	5.7	41.8
Subei	26.0	10.5	40.3	9.5	36.4	6.1	23.3
Xuzhou	9.6	3.3	34.6	4.1	42.4	2.2	22.9
Huaiyin	4.5	2.5	55.6	1.1	24.5	0.9	19.9
Liangyungan	5.1	2.1	42.0	1.7	33.4	1.2	24.6
Yancheng	6.8	2.5	36.8	2.6	38.2	1.7	25.0

Source: JSB 1996.

Post-Mao economic reforms have provided favorable conditions for the development of non-state enterprises, including TVEs. The share of industrial output produced by SOEs has declined further (Fig. 6.4). Among policies benefiting TVEs were low tax rates, deregulation of prices, budget bonuses for growth, and new labor policies. As mentioned earlier, Sunan had a larger proportion of non-state enterprises to begin with, and therefore quickly capitalized on the reform policies which favored non-state sectors. On the other hand, regions with large proportions of SOEs, Subei in particular, have grown slower because SOEs have had difficulty in adjusting to the emerging market mechanisms. In 1995, investment in SOEs accounted for a half of fixed asset investment in Huaiyin (Subei), compared with only 19.6 percent and 16 percent in Suzhou and Wuxi (both in Sunan), respectively. As a result, Subei has lagged behind in economic growth. By the same token, the concentration of SOEs in the cities has contributed to the slower growth of cities than counties that depend largely on TVEs.

Besides fiscal and investment reforms, the provincial government of Jiangsu has promoted local economic growth in many other aspects. Provincial officials have actively bargained with the central government to obtain more favorable open door policies for localities. All these factors provided incentives for local leaders to initiate new policies for the growth of TVEs and to engage in the production process. For example, township governments in Suzhou introduced a contract responsibility system to Suzhou's township enterprises as early as 1979.[6] The efforts of provincial and local governments are important to the development of TVEs.

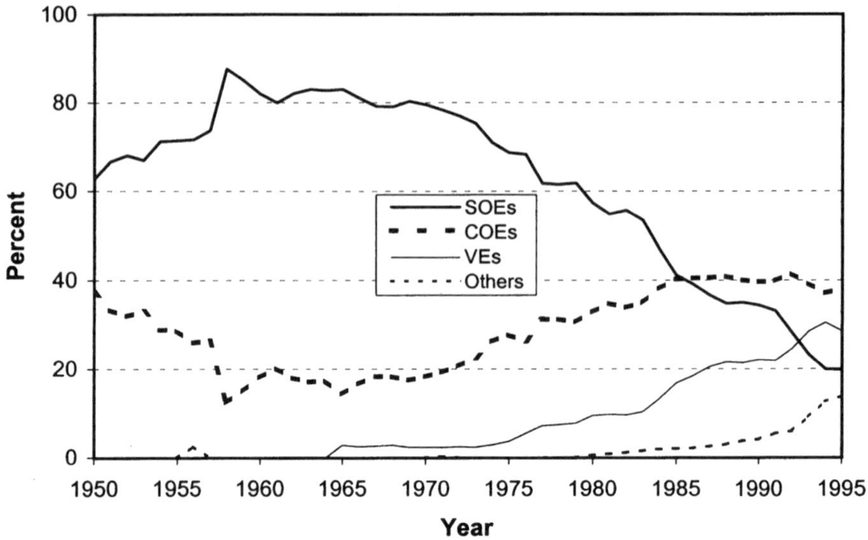

Figure 6.4 Changing shares of industrial output produced by different forms of ownership, 1950–95.
Source: JSB 1996.

The development of TVEs and Sunan also benefit from agglomeration economies, especially linkages to the nearby cities. The economic ties to Shanghai have been particularly important to the development of Sunan. Shanghai was seceded from Jiangsu in 1927 to become a provincial-level administrative unit. In 1958, 10 counties from Sunan were incorporated by the Shanghai administration. Jiangsu has had historically close ties with Shanghai, and TVEs in Jiangsu provide goods for the consumers and enterprises in Shanghai. During Mao's era, numerous youths and intellectuals in Shanghai were 'rusticated' to Jiangsu, particularly to Sunan (Du and Gu 1987). They became the primary source of technical and management know-how for the initial development of TVEs in Jiangsu, and forged production and marketing ties between Shanghai and Jiangsu. Some of them still reside in Sunan, as they were not permitted to return to Shanghai due to tight population control (Jacobs and Hong 1994).[7] During the 1980s, 80 percent of enterprises in Taicang county had cooperative projects with Shanghai (Jin and Zhu 1993). In 1991, 27.8 percent of input procurement and 39.4 percent of markets for rural enterprises in Kunshan were with Shanghai (Marton 1995).

Besides Shanghai, TVEs have established close ties with many other cities, particularly major cities in Sunan (Suzhou, Wuxi, and Changzhou). The provincial government and local governments in Sunan encouraged TVEs to strengthen their ties with these cities. The new administrative system, the 'city-leading-county' system, has been implemented to improve the cooperation between the cities and counties. Numerous delegates have been sent to Shanghai and other cities to improve economic cooperation. During the Sixth FYP (1981–85), Jiangsu had 3500 joint economic projects with 28 provinces (Editorial Committee 1989). As early as in 1985, 80 percent of enterprises in Yangzhong (in Suzhong) had already established numerous cooperative projects with 1400 economic and research units all over the nation (JPERC 1988). About 70 percent of TVEs in Sunan have established cooperative projects with enterprises and research institutes in the cities. Cities have brought increasing wealth to adjacent rural areas (Veeck and Pannell 1989). Therefore, the development of Sunan must be explained in relation to the ties between rural areas and the cities, and between Jiangsu and Shanghai.

Development paths of the regions in Jiangsu have important implications for their growth patterns. During Mao's era, the central, provincial, and local governments stressed agricultural production in Subei, and few resources were committed to the development of light industry and TVEs. While Sunan grew rapidly through industrialization, Subei lagged behind due to the dominance of slowly growing agriculture and SOEs. It was also earmarked in the 1950s for the development of heavy industry, particularly coal, energy, and heavy machines. Sunan, however, was traditionally dominated by light industry with a large proportion of collective enterprises. From 1978 to 1995, output of SOEs in Jiangsu increased from 20.8 billion yuan to 249.8 billion yuan, an increase of 1101 percent, much slower than the growth of COEs (4460 percent) (current price).

While Suzhong has made some progress in rural industrialization, Subei is still dominated by slowly growing agricultural production and SOEs. In 1995, Subei and the old cities had higher percentages of industrial output produced by SOEs than Sunan (Fig. 6.5). As shown in Table 6.6, per capita output of township enterprises (TEs) in Sunan was ten times higher than that in Subei. In addition, in 1994, the share of the primary sector (agriculture) in the total output was 7.2 percent, 13.7 percent, and 32.5 percent for Sunan, Suzhong, and Subei, respectively. In Subei and Suzhong, more than 40 percent of investment was located in less profitable SOEs, while in Sunan, more profitable TVEs and FIEs absorbed the bulk of the investment. The large shares of SOEs have hindered the development of Subei. Agricultural production and SOEs have stagnated in Subei, while Sunan has experienced rapid growth through the development of non-state sectors.

Regional resource endowments are also important to the development of TVEs. TVEs have been less supported by the central government and have to rely on local resources more than SOEs. Sunan has long been one of the most productive agricultural bases in China, and has provided capital and raw materials for the development of TVEs. Moreover, due to the lack of state investment, TVEs rely heavily on local resources to finance construction and production. Many activities of processing agricultural products were

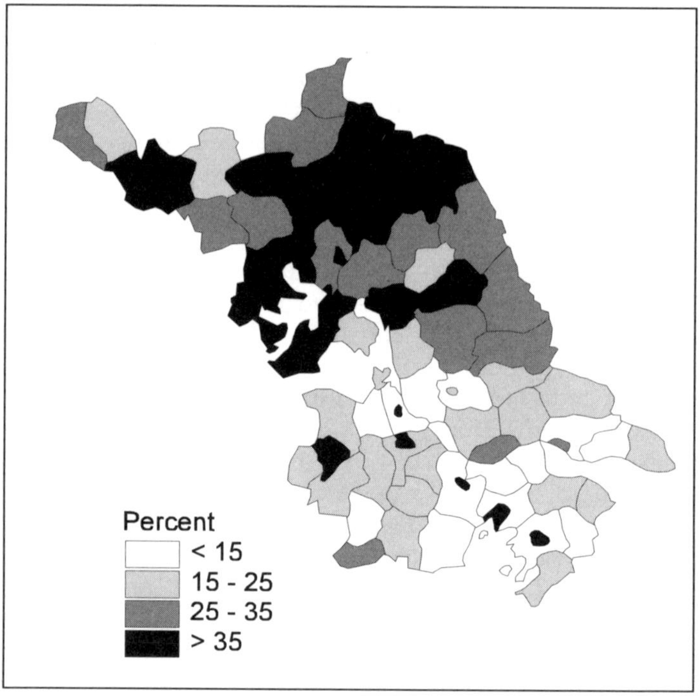

Figure 6.5 Regional disparities in shares of industrial output produced by SOEs, 1995.
Source: JSB 1996.

traditionally dispersed throughout the countryside (Shih 1992). These advantages help TVEs develop strong abilities to adjust themselves to the constantly changing macro economic policies and markets.

The development of TVEs has important implications for economic and regional development in Jiangsu. In 1995, TVEs produced 706 billion yuan industrial output, which accounted for 56 percent of the gross value of industrial output (GVIO) in Jiangsu. TVEs have clearly become the most important form of industrial ownership in Jiangsu—more important than SOEs, which only produced 19.8 percent of GVIO. Moreover, as shown in Table 6.7, SOEs tend to be less profitable than COEs, as COEs have a higher capital–output ratio and less loss-making enterprises. The financial conditions of SOEs were even worse in 1997, as SOEs lost 3.07 billion yuan, and 45.2 percent of SOEs lost money (Song *et al.* 1998). Among Jiangsu's 1097 large and medium-sized SOEs, 409 enterprises, or 37.28 percent, lost money. While many provinces in China have been troubled by low-efficiency SOEs,

Table 6.6 Regional output of township enterprises (TEs), 1995

Region	*Volume (billion)*	*Share (percent)*	*Per capita (yuan)*
Sunan	178.2	31.8	13,364
Suzhou	75.3	25.8	13,160
Wuxi	74.5	40.8	17,376
Changzhou	28.4	32.9	8,540
Suzhong	85.5	24.6	3,414
Nanjing	15.6	15.0	3,000
Zhenjiang	15.3	27.7	5,821
Yangzhou	30.6	27.2	3,263
Nantong	4.0	31.3	3,064
Subei	37.6	21.8	1,248
Huaiyin	7.0	19.9	682
Lianyungang	4.3	17.7	1,201
Xuzhou	11.3	18.6	1,334
Yancheng	15.0	28.8	1,918

Source: JSB 1996.

Table 6.7 Major efficiency indicators of SOEs and COEs, 1995

Indicator	*SOEs*	*COEs*
Ratio of profits and taxes to fixed assets	9.2	19.8
Percent of loss-making enterprises	30.6	21.9

Source: JSB 1996.

Jiangsu, Sunan in particular, with less SOEs, has performed better in economic growth.

The significance of non-state enterprises in the development of Jiangsu is more than a contribution to output and efficiency. TVEs are the major sources of employment and taxes for townships and villages in Jiangsu. The development of TVEs has stimulated industrialization and urbanization, making Jiangsu one of the most urbanized provinces in China. The structural difference in industrial ownership and efficiency has contributed significantly to the gap between Sunan and Subei, and the problems with the old cities where much of the SOEs are concentrated.

While TVEs have played a significant role in the economic and regional development of Jiangsu, they are at the crossroads for future development. The nature and property rights of TVEs have never been clear, and with the deepening of economic reforms, such a form of ownership has created some problems in property rights, income distribution, and party–enterprise relationships. In addition, TVEs tend to be small in size and lack economies of scale and agglomeration. Their labor force tends to be less trained, and they are less capable of technical innovation and improving management requirements than modern corporations. In 1993, reforms of the ownership structure of TVEs were initiated in some rural areas, and have been accelerated since 1996. Alternative forms of ownership have been introduced into TVEs, including shareholding systems and joint ownership with foreign investors and private investors. TVEs are preparing themselves for intense competition from FIEs, successfully transformed SOEs, and emerging private enterprises.

6.4 Globalization, foreign investment, and regions

Globalization and China's reforms have increased the importance of global forces in economic and regional development in China. The Chinese government has provided preferential policies in taxation, infrastructure improvement, tariffs, and currency exchange to attract foreign investment. As I have explained previously, China's efforts to attract foreign investment have been facilitated by the globalization of economic activities during which an increasing amount of foreign investment has flowed to developing countries to take advantage of cheap labor and large potential markets. Moreover, economic restructuring in Asian countries, particularly in the Asian NICs, has also facilitated the dramatic increase of foreign investment in China.[8]

As I have shown in Chapter 4, China's open door policy is spatially uneven, biased towards the coastal cities. Open zones and cities have enjoyed more preferential policies and greater autonomy in terms of foreign investment, trade, and currency exchange. FDI is concentrated in China's coastal provinces as global investors can receive higher investment returns there largely due to favorable state policy and local conditions. FDI is also unevenly distributed within provinces. Jiangsu is one of the dynamic coastal provinces that have enjoyed preferential open door policies and have made considerable efforts to attract foreign investment.

6.4.1 State policy and FDI development

Jiangsu started later than Guangdong and Fujian in opening up to the outside world, as China's open door policy during the early 1980s focused on SEZs located in Guangdong and Fujian. In Jiangsu, two agreements with foreign investors were approved in 1981, and these projects began operation in 1983. In 1982 and 1983 one agreement was approved in each year, but neither was operational before the end of 1985 (Pomfret 1989). In 1984, China's open door policy was expanded to other coastal provinces, including Jiangsu. Nantong and Lianyungang were among China's 14 open coastal cities designated to attract foreign investment and promote foreign trade. In 1985, all cities and counties in Sunan, including Suzhou, Wuxi, Changzhou, and 12 counties under their jurisdiction, became open areas.

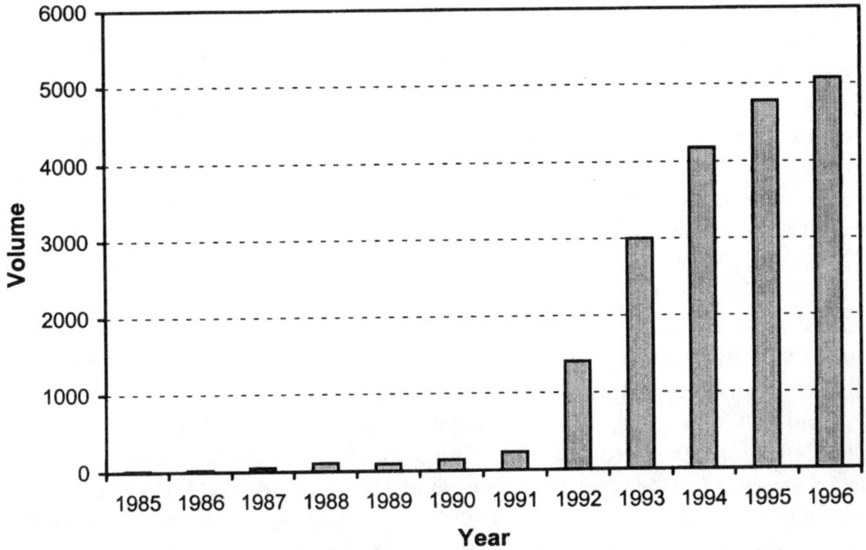

Figure 6.6 The growth of FDI in Jiangsu, 1985–96.
Source: JSB 1991–97.

Their local governments obtained preferential policies in the approval of investment projects, tax relief for using foreign capital, tariff reduction on imports, and the retaining of exchange earnings. Most of the open areas are located in Sunan, while only the city of Lianyungang is located in Subei and the city of Nantong in Suzhong. FDI in Jiangsu was small in size (Fig. 6.6).

Major steps were taken in 1988 to accelerate foreign investment and trade, and Nanjing and Zhenjiang joined the list of open cities. In total, nine municipalities and 40 counties in Jiangsu were opened up for foreign investment and trade. These cities received provincial-level authority in approving foreign investment projects, and foreign-invested projects could be approved locally if their value was less than US$5 million (Pomfret 1989). Joint ventures (JVs) enjoyed privileges in taxes, local contents, forms of investment, exports, site preparation, and many other aspects. While areas in Sunan and along the coast became open areas, Xuzhou and Huaiyin, two municipalities in Subei, were still not opened up for foreign investment and trade. Most of the areas opened up in this year were in Suzhong, and Suzhong therefore benefited greatly from the 1988 open door policy. This year, the first wholly foreign-owned enterprise (WFOE) was established in Changzhou, a city located in Sunan. In 1989, FDI in Jiangsu staggered, as a result of the Tiananmen incident and economic stagnation. In 1990, the provincial government decided to accelerate absorbing FDI with more favorable policies, including authorizing city governments the power to approve projects with a value less than US$30 million and simplifying project approval procedures.

A series of reform and open door policies were introduced in the early 1990s, including policies specifically targeting investors from Hong Kong, Taiwan, and Singapore, which, as well as Japan, are the major sources of FDI in Jiangsu. Numerous open districts have been established to attract foreign capital: the largest and most successful ones are all located in Sunan. Indeed, government policies in Jiangsu were more generous than anywhere in China outside SEZs (Pomfret 1995), and after 1991, FDI in Jiangsu increased greatly (Fig. 6.6). During the mid-1990s, following the change of national regulations, Jiangsu made some adjustment in its open door policy, including classifying investment into encouraged (e.g., high-tech industry, infrastructure, energy and raw materials) and restricted categories (those generating pollution and consuming energy).

From 1985 to 1996, FDI in Jiangsu increased from US$11.9 million to US$5.1 billion. Its share of FDI in China increased from 1.5 percent in 1983 to 4.3 percent in 1990 and 13.9 percent in 1995 (the second highest in China, only less than Guangdong). From 1978 to 1997, Jiangsu has approved 34,226 foreign-invested projects and absorbed US$27.1 billion in FDI, only behind Guangdong. Equity joint ventures (EJVs) took much of FDI during the early years of the reform period (e.g., 95.7 percent in 1985), but investment in WFOEs has been rising rapidly. In 1995, EJVs accounted for 69.5 percent of FDI, followed by WFOEs (24.7 percent) (JSB 1997b). The increasing importance of WFOEs, as a part of the national trend, represents the increasing confidence of foreign investors in managing enterprises in a different cultural and political environment, and the release of government control over foreign ownership of enterprises in China.

Asia has been the major source of FDI in Jiangsu, followed by North America and Europe. In 1995, FDI from Hong Kong reached US$1.6 billion, which accounted for 34.1 percent of FDI in Jiangsu, followed by Japan (US$641 million), Singapore (US$575 million), Taiwan (US$492 million), and the United States (US$365 million). In 1996, Jiangsu was the largest host province in China for investment from Taiwan, although source countries of investment change yearly. In terms of industrial allocation, the manufacturing industry took in much of FDI in Jiangsu (82.6 percent in 1995). In 1995, major industries attracting FDI included textiles (US$435 million), real estate (US$358 million), electronics and communication equipment (US$335 million), general manufacturing (US$303 million), and clothes and fiber products (US$291 million) (JSB 1997b).

6.4.2 Uneven regional distribution

FDI is unevenly distributed in Jiangsu. The gap among Sunan, Suzhong, and Subei was large. In 1995, 72.1 percent of FDI in Jiangsu was concentrated in Sunan; Suzhou alone attracted 48.7 percent of FDI (Table 6.8). Total amount of

Table 6.8 Regional distribution of foreign direct investment, 1995 (US$)

	Volume (million)	*Share (percent)*	*Per capita*
Jiangsu	4,780.6	100.0	67.9
Sunan	3,446.5	72.1	258.5
Suzhou	2,327.5	48.7	406.8
Wuxi	857.4	17.9	200.0
Changzhou	261.7	5.5	78.7
Suzhong	1,033.9	21.6	41.3
Nanjing	402.8	8.4	77.5
Zhenjiang	164.8	3.4	62.7
Yangzhou	158.7	3.3	16.9
Nantong	307.6	6.4	39.3
Subei	273.4	5.7	9.1
Xuzhou	104.3	2.2	12.3
Huaiyin	38.7	0.8	3.8
Lianyungang	69.9	1.5	19.5
Yancheng	60.5	1.3	7.7

Source: JSB 1996.

and per capita FDI in Subei were considerably less than Sunan, due to less favorable state policy and local conditions. These patterns of foreign investment have further intensified regional concentration of domestic capital (Su and Veeck 1995).

Figure 6.7 clearly shows the concentration of FDI in Sunan. In 1995, among counties, Zhangjiagang attracted US$599.8 million FDI, followed by Kunshan (US$505.3 million), Jiangyin (US$252.3 million), and Wuxian (US$251 million); all are located in Sunan. Several counties in Subei, including Fengxian, Lianshui, and Sihong, had FDI less than US$0.5 million. Large gaps also exist among cities in attracting foreign investment, with FDI in Suzhou (US$558.5 million) much larger than that in Huaiyin (US$16.2 million).

Most of the open development districts are concentrated in Sunan. In the mid-1990s, Jiangsu established 11 national-level and 63 provincial-level development districts. These development districts provide well-developed infrastructure and services to foreign investors. In 1997, these districts absorbed US$2.76 billion FDI, which accounted for 55.5 percent of the provincial total (People's Government of Jiangsu Province 1997). There were two development districts with output surpassing 10 billion yuan and seven development districts having output of more than 6 billion yuan. Most of the largest development districts attracting FDI are located in Sunan, including Suzhou Industrial Park, Kunshan Economic and Technological Development District, Zhangjiagang Tariff-Free District, Nanjing High-Tech Development District, and Wuxi Industrial Park. In 1996, Suzhou Industrial Park absorbed US$410 million FDI, much larger than the development districts in Subei, such as Liangyungang Development District.[9]

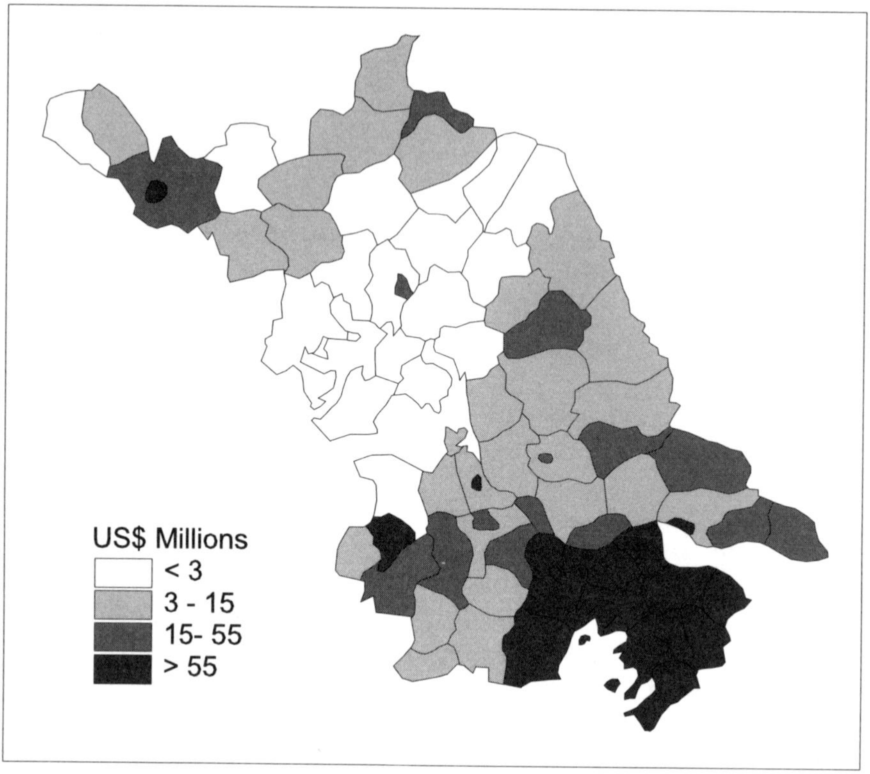

Figure 6.7 Foreign direct investment in Jiangsu, 1995.
Source: JSB 1996.

The uneven geographic distribution of FDI is related to global, national, and local factors. First, reform and open door policies have been biased towards Sunan in general: most favorable policies and the largest open districts are all located in Sunan. This regional bias of FDI is consistent with the national pattern in which FDI tends to be concentrated in more developed coastal provinces. This supports the belief that developing countries, in their efforts to attract foreign investment, tend to provide more favorable policies to the regions which are more effective in attracting FDI.

Secondly, local governments in Sunan have been more active in attracting FDI, and Sunan also provides better industrial linkages, markets, labor, and infrastructure for foreign investors. As I have explained previously, local governments in Jiangsu are more active in initiating local development policies than many other provinces of China. However, as I will show in the case of Kunshan, government policies are also related to local development conditions. Sunan has advantages in terms of local states and local development conditions, and is more attractive to foreign investors.

Lastly, Jiangsu has benefited from production outsourcing in developed countries and Asian NICs. Jiangsu has long established close cultural and economic ties with foreign countries, including Taiwan and Hong Kong. In short, Sunan's established networks, favorable location and resources, and efforts of local governments have helped the dramatic growth of FDI in Sunan and produced the uneven geographic distribution of FDI in Jiangsu.

Foreign trade has also grown rapidly in Jiangsu. From 1978 to 1996, exports increased from US$418 million to US$13.1 billion, an annual increase of 21.1 percent (JSB 1997b). The growth rate during the Eighth FYP period (1991–95) was 31.9 percent, considerably larger than that in the Sixth FYP period (1981–85) (12.8 percent) (JSB 1997b). In 1996, major export partners included Japan (US$4.1 billion), Hong Kong (US$2.6 billion), and the United States (US$2.1 billion). Like foreign investment, Sunan also dominates foreign trade. In 1995, exports from Sunan were US$5.8 billion, which accounted for 59 percent of the total exports; Suzhou's exports were US$3.2 billion, which accounted for 35.9 percent of the total (JSB 1997b).

The concentration of FDI in Sunan has provided abundant capital for its economy and has stimulated the rapid growth of Sunan, while Subei has lagged far behind. In 1996, FDI accounted for 23 percent of total fixed investment in Jiangsu (People's Government of Jiangsu Province 1997). Increasing foreign investment has also contributed to the rapid growth of foreign trade in Jiangsu. In 1996, 'self' export of FIEs reached US$6.6 billion, which accounted for 43.3 percent of total 'self' export of enterprises (Song *et al.* 1998). FIEs provided more than 2 million jobs, and greatly enhanced technical level, management skills, and labor quality in Jiangsu. However, the contribution of FDI to economic development is uneven across regions in Jiangsu, as Sunan has benefited more from the open door policy and the increase of FDI. Subei is still lagging far behind in attracting FDI and promoting exports.

6.5 Beyond the Sunan model: development of Kunshan

Closely related to this research is the Sunan model of development, which attributes the rapid economic growth of Sunan to the development of TVEs, and is often conceptualized as 'development from below.' This conceptualization, however, needs to be improved to account for the development of Sunan, as the role of macro economic reform policies, the importance of Shanghai and major cities in Sunan, and the influence of foreign investment and trade should also be emphasized in Sunan's economic growth. In addition, the role of local states in development would be handicapped without proper macro, local, and external conditions for regional development. Regional development cannot be simply understood as a state-centered or export promotion process. Rather, as I have been arguing in this book, the functioning of the central state, local states and agents, and global forces

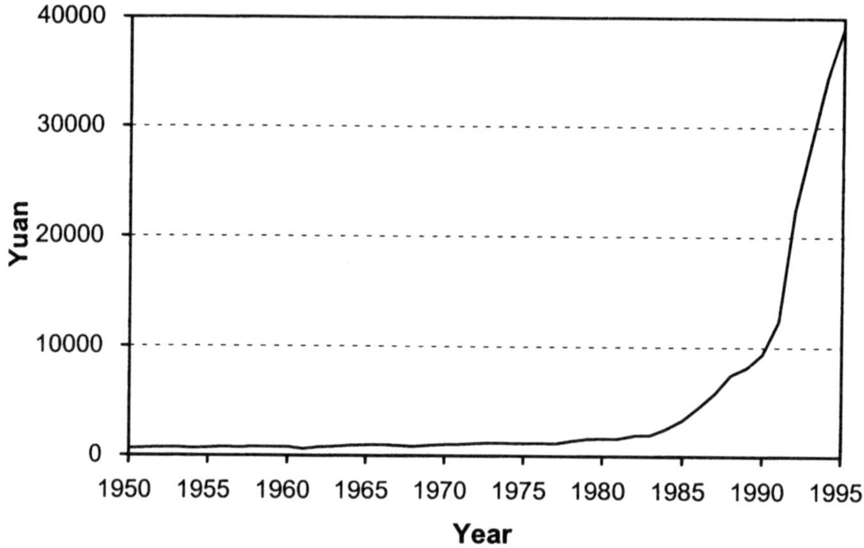

Figure 6.8 Gross industrial and agricultural output in Kunshan, 1950–95.
Source: JSB 1994b; 1994–96.

underlies regional development, which can be elaborated through a case study of Kunshan, a county-level city. Kunshan is located in Sunan, and is adjacent to Shanghai (Fig. 5.1).

Kunshan's economic performance under Mao was poor, and per capita output was stagnant (Fig. 6.8). From 1950 to 1978, its annual average growth rate of per capita GVIAO was only 3 percent, lower than the average of Jiangsu (4.9 percent) (Table 6.9), and considerably slower than cities that were favored by Mao's industrialization policy. While Mao in certain periods, such as the late 1950s and early 1970s, encouraged decentralization and rural development which laid the foundation for the development of TVEs, the effects of such policies could not offset the city-centered, heavy industry-oriented Soviet model of development. Kunshan's slower growth resulted in the decline of its status in the economy of Jiangsu, as evidenced by the decrease of location quotient (LQ) from 185 (higher than the provincial average) in 1950 to 97 (slightly lower than the average) in 1977 (Fig. 5.14). In 1977, per capita GVIAO in Kunshan was only 1095 yuan, among the lowest of the counties in Sunan, although it was still higher than poorer Subei. Little effort was made by Kunshan to develop TVEs, and its economy was dominated by agriculture, which produced 61.6 percent of GVIAO, much higher than the average level in Suzhou municipality (33.6 percent) (JSB 1994b). Per capita income of rural peasants was 131 yuan, only slightly higher than that in 1956 (96 yuan), with an annual growth rate of 1.5 percent (KSB 1993).

Table 6.9 Major indicators in Kunshan and Jiangsu province

Indicator	*Kunshan*	*Jiangsu*
Per capita GVIAO (1990 constant price)		
1950 (yuan)	615	334
1978 (yuan)	1,340	1,291
1995 (yuan)	31,280	11,129
1950–78 Average annual growth (percent)	3.0	4.9
1978–95 Average annual growth (percent)	20.4	13.5
1995 population (million)	0.58	70.66
1995 Per capita		
Fixed asset investment (yuan)	7,748	2,378
Foreign direct investment (US$)	871	75
Revenue	881	495

GVIAO: gross value of industrial and agricultural output.
Source: JSB 1994b; 1996.

During the late 1970s and the early 1980s, periods which emphasized rural reforms, Kunshan's status in the provincial economy was not improved, as rural reforms mainly improved agricultural production and rural development in Subei. Meanwhile, in Sunan, TVEs began to enter a stage of rapid growth, although TVEs' early development can be traced back to the late 1950s and early 1970s. Kunshan, however, did not enjoy the opportunities created for agricultural development and rural industrialization. As a result, by 1983, Kunshan's LQ was only 98, still lower than the provincial average and among the lowest in the counties in Sunan.

With the shift of the emphasis to urban and industrial reforms in 1984, the government of Kunshan initiated economic and regional planning which set up ambitious development objectives. A major strategy was the decision to establish 'Kunshan Economic and Technological Development District' (KETDD, or the development district) in the eastside of Yushan town. Indeed, the year 1984 marked the beginning of a period of phenomenal growth. In 1985, Kunshan was among the counties in Jiangsu, and in China, that were opened up the earliest for foreign investment and trade. In 1989, it was designated as a county-level city due to significant achievement in urban and economic growth. During the 1978–95 period, average annual growth rate of per capita GVIAO was 20.4 percent, one of the highest among Jiangsu's counties. By all means, Kunshan's growth since the mid-1980s has been spectacular, and it has become one of the richest counties in China. In 1995, Kunshan was ranked the 9th in comprehensive economic strength among all of the counties in China (Zhang 1996). In 1996, per capita GDP reached 19,656 yuan, ranked the 4th in Jiangsu. Rapid growth has also been accompanied by the development of the countryside, as per capita income of peasants reached 4101 yuan in 1996. This record of growth, however, cannot simply be attributed to the development of TVEs (the Sunan model) or to export-oriented development. Kunshan's development is very much a 'mixed model', incorporating elements of state intervention, utilization of local resources, and outward orientation: i.e. the functioning of global, state, and local forces, as argued in this book.

The central, provincial, and local governments have played a significant role in the development of Kunshan. The central government has provided favorable macro conditions for Kunshan's economic growth. Kunshan, like other coastal counties, can keep portions of its growing revenues and has received considerable power in resource allocation. Its development district was designated by the central government as China's first national-level opening district of counties that enjoy tremendous autonomy. The provincial government of Jiangsu has actively bargained with the central government to obtain more favorable reform policies for the counties. It has also helped regions to improve interregional networking and facilitated economic reforms by promoting successful reform experiences.

The government of Kunshan has played a critical role in economic development. First, Kunshan leads the province in initiating numerous local policies to stimulate economic development, including financial reform, labor reform, and open door policy. It has also been keen in reforming social institutions in order to promote economic growth. Secondly, it acts as an agent in accumulating capital and negotiating regional and global investments. One of the most important branches of the Kunshan government is the KETDD Administration Commission that manages more than 20 state enterprises. The government has helped to establish extensive networks with numerous cities in China to secure raw materials, technology, and capital. Thirdly, local officials act as entrepreneurs in creating and managing enterprises. While the central and provincial governments are responsible for creating macro- and meso-level environments for economic growth, the local state in Kunshan, like many others in Jiangsu, acts both as facilitator and engineer (director). As a facilitator or a development state, it strives to create an environment for the development of state, collective, and foreign-invested enterprises. As a 'director,' it acts very much like an entrepreneurial state that directly leads economic development and manages industrial production.

The functioning of the local state, however, is embedded in local development conditions. Local conditions strongly influence the strategy and daily operation of local governments. Kunshan enjoys advantages in location, infrastructure, spatial networks, and development experiences. Kunshan is located in the Tai Lake (Taihu) region, which provides a favorable setting for fisheries and agriculture. It is the most prolific agricultural region and one of the richest places in China, in which urban and rural areas have long established close ties. Such historically established linkages have been strengthened by the Kunshan government, which has sent numerous delegates to Shanghai and other cities to improve economic cooperation. Transportation has also been improved by recent highways and freeways development. Such favorable conditions in location and transportation are hardly matched by other places in Jiangsu and in China, and lay the foundation for close economic linkages with the cities. As early as 1985, 60 percent of factories in Kunshan had already established certain kinds of production relations with enterprises located in the cities (Zhang 1994). The most important linkages are with Shanghai and with large cities in Sunan. As Yushan is only 55 km from Shanghai and 37 km from Suzhou, Kunshan is very much like a suburban county of both Shanghai and Suzhou. Naturally, it became a favorable location for outsourcing Shanghai's production and industrial relocation. Through these networks, Kunshan effectively attracted domestic and international investment. Therefore, its development must be explained in terms of its local development conditions and its relation to the ties between cities and regions, and between Kunshan and Shanghai. While some of the local conditions prevail in many regions of China, others are historically and geographically specific to Kunshan.

Kunshan has recorded a dramatic growth in foreign investment, which also contributes greatly to its phenomenal growth. In 1997, FDI in Kunshan reached US$560 million (the second highest among counties in Jiangsu), and exports were US$1.1 billion, of which 93.4 percent were by FIEs (KSB 1998b). In 1997, FIEs controlled 9.72 billion yuan in fixed assets, which accounted for 69.1 percent of total fixed assets in Kunshan (KSB 1998b). Besides favorable open door policies to coastal areas, the local state and locational advantages are important to the growth of foreign investment and trade in Kunshan. In the mid-1980s, the development district (i.e. KETDD), started with an area of 3.75 km^2, but was soon enlarged to 6.18 km^2. To make the district attractive to investors, the government has made great efforts to improve the infrastructure, including transportation, electricity, communications, water supply, sewerage, gas, and land conditions. The government also emphasizes efficiency in the investment approval process and provides administrative and financial supports for foreign investors. Kunshan has been especially successful in attracting investment from Asian NICs, including Taiwanese investors, as the provincial capital, Nanjing, was the former national capital of the Republic of China (Taiwan). Such historically established social networks and cultural affiliation helped to bridge Taiwan investors with Jiangsu and to make Kunshan a favorable host location of Taiwanese investment. In summary, the central state, local efforts and factors, and global restructuring have all contributed to the emergence of foreign investment and trade in Kunshan, which in turn shapes the development of Kunshan.

The Kunshan experience shows that regional development is a complicated process that incorporates forces operating at the global, national, and local levels. Oversimplified conceptualization of the development of Sunan, as described by the Sunan model, masks varied development paths taken by cities and counties in the region and other important factors that shape the development process. The Kunshan model, in a certain way, resembles the development process of Asian NICs, but with stronger local states and different global and national contexts. The Kunshan experience shows that:

1 An active state can spur growth and development.
2 Foreign investors, in choosing locations for investment, tend to consider government policies, local development conditions, as well as their own needs.
3 Local conditions significantly contribute to local development, and are the basis for the formation of government policies and the locational choice of foreign investment.

In other words, global, state, and local forces have all contributed to the emergence of Kunshan and uneven development in Jiangsu.

6.6 Understanding regional inequality in Jiangsu

Previous sections have examined changing patterns and underlying forces of regional development and inequality in Jiangsu. I have highlighted the state, local states and development conditions, and foreign investment as shaping uneven regional development. First, the Maoist period stressed urban industrialization, which stimulated the growth of cities, but it also laid the foundation for the development of TVEs through decentralization and rural industrialization policies, especially during the early 1970s. Within Jiangsu, agriculture in Subei was emphasized; this grew much slower than industry in Sunan. Also in Subei, considerable resources were allocated to the development of heavy industry, which was less efficient than light industry. Post-Mao economic reforms have provided more favorable policies for the development of Sunan, including the open door policy, policies to adjust economic structure, economic decentralization, and reforms of financial and investment systems. TVEs, largely located in Sunan, enjoy lower tax rates and more autonomous decision making. In short, state policy has provided favorable conditions for the development of Sunan, both during Mao's era and the reform period.

Secondly, local states and development conditions are important to regional development. Sunan has long been one of the most developed regions of China. Local governments in Sunan have acted as development/entrepreneur states that actively promote and seek resources for local economic development. The growth of Sunan, especially its TVEs, also relates to its close socioeconomic ties to nearby cities. Shanghai, in particular, has provided a large market and strong technical support for the development of Sunan. Moreover, Sunan's strong agricultural base, well-established transportation systems, and a well-educated labor force have all contributed to its economic prosperity. Subei and Suzhong, however, are in a disadvantaged position in terms of local factors of regional development.

Lastly, foreign investment and trade have also contributed to the growth of Sunan. It was opened up for foreign investment earlier than Subei, and has enjoyed more favorable open door policies provided by the central government. Local governments in Sunan have also actively sought foreign capital for local economic growth. Moreover, Sunan has better local conditions in labor, transportation, and resources, which are attractive to foreign investment. FDI in Jiangsu is concentrated in Sunan, largely as a result of the regional bias of state policies and favorable local factors for foreign investment. In short, the state, local states and conditions, and foreign investors have all contributed to the rapid growth of Sunan and the rise of interregional inequality in Jiangsu. Especially with the shift of reforms to urban/industrial reforms and open door policies in the mid-1980s, the Sunan–Subei gap has been intensified.

While interregional inequality has increased, changes of (overall) intercounty and rural (intercounty) inequalities are more complicated.[10] Contrary to the normal expectation that Mao's policy reduced income inequality, intercounty and rural inequalities in Jiangsu increased under Mao. Although some efforts were made to develop backward counties, major cities recorded faster growth, due largely to Mao's industrialization policy, which favored the growth of cities. On the other hand, local initiatives and Mao's decentralization policy to a certain extent facilitated the growth of Sunan. Mainly as a result of faster growing cities, intercounty inequality rose under Mao, rather than declined.

Post-Mao economic reforms have stimulated the rapid growth of counties in Sunan. Local efforts and the development of TVEs have stimulated the rapid growth of these counties. In addition, government policies and foreign investment have also contributed to the development of non-state sectors and counties in Sunan. However, major cities, which had the highest industrial output and recorded faster growth under Mao, experienced slower growth. Strict state control, the failure of reforming SOEs, and overburdened infrastructure have contributed to the slower growth of the leading cities. The slow growth of cities and rapid growth of counties resulted in a decline of intercounty inequality during the reform period, which is not the result of any efforts by the governments to reduce regional inequality. Rural inequality, however, has risen since reforms, especially since the mid-1980s and early 1990s, due largely to the rapid growth of counties in Sunan.

During the 1990s, the richest counties in Sunan registered much faster growth. More intense reforms for decentralization, marketization, and globalization have created favorable conditions for the development of non-state sectors. Meanwhile, foreign investment also increased dramatically in Sunan. These factors have contributed to the rapid growth of counties in Sunan. By 1995, the industrial and agricultural output of several counties had actually surpassed that of the cities. The emergence of these rich counties in Sunan and the slow growth of its cities have contributed to a renewed increase of intercounty inequality in Jiangsu after the mid-1990s, although the rise of rural inequality lessened.

Table 6.10 Regression analysis for uneven growth across Jiangsu's cities and counties

Independent variable	*Standard coefficient*	*Regression error*	*Standard significance coefficient*	*T-value*	*Level (P)*
FAI	0.115	0.043	0.336	3.622	0.001
SOE	−3.722	1.682	−0.217	−2.213	0.030
FDI	2.852	0.766	0.325	3.722	0.000
CITY	−3.766	1.090	−0.307	−3.456	0.001
Constant	11.297	3.473	0.0	3.252	0.002

Dependent variable GVIAOPG
$R = 0.794$ $R^2 = 0.63$ *Adjusted* $R^2 = 0.610$
$F = 29.882$ *Significance level (P)* $= 0.000$

Note: Dependent variable GVIAOPG: average annual growth rates of GVIAO per capita (percent), 1978–95. Independent variables: (1) FAI: average annual growth rate of fixed asset investment per capita (percent), 1978–95, (2) SOE: share of fixed asset investment in state-owned enterprises (percent), 1995 (transformed using logarithms); (3) FDI: ratio of foreign direct investment to fixed asset investment (percent), 1995 (transformed using logarithms); (4) CITY: a dummy variable representing the cities.

I have performed regression analysis to examine the contribution of state, local factors, and foreign investment to uneven regional growth. The dependent variable is the growth rate of GVIAO per capita (GVIAOPG) (Table 6.10). The units of analysis are counties (including county-level cities) and cities (*shiqu*). The period of analysis is from 1978, the year when economic reforms were launched, to 1995, the end year of this research. Based on the analysis of global, state, and local forces, I have selected the following four independent variables with a focus on investment input:

1 FAI is expected to contribute positively to regional growth (Table 6.10).
2 SOE represents ownership structure and is expected to be negatively related to regional GVIAO per capita (GVIAOP) growth.
3 FDI has been widely used to represent the importance of global forces and should contribute positively to regional economic growth.
4 A dummy variable (CITY) is assigned to cities that have been slow in adjusting to economic reforms.

The results of regression analysis are reported in Table 6.10. The regression equation as a whole and the regression coefficients are all significant at the 5 percent level. R^2 is 0.631, indicating that 63.1 percent of the variation in regional GVIAOP growth rate is explained by the independent variables. GVIAOPG is positively related to FAI and FDI, indicating that regions with faster investment growth and larger shares of foreign investment tend to grow faster. GVIAOPG is negatively related to SOE and CITY, indicating that regions that have more significant presence of state-owned enterprises and are cities in location tend to lag behind. In terms of the relative importance of the independent variables, investment growth (FAI) is the most important, followed by FDI, CITY, and SOE, as indicated by standard regression coefficients. Regression analysis suggests that state policy remains important in regional development, but that emerging local and global forces are becoming important. It strongly supports my arguments for the importance of state policy, localities, and foreign investment in Jiangsu's regional development.

6.7 Conclusion

This chapter has shown that three sets of mechanisms—state policy, local states and development conditions, and foreign investment—are important to the understanding of regional development in Jiangsu. While research in Jiangsu tends to emphasize forces operating at the local level, state policy and foreign investment are also exerting great influences. A synthesis of the three major agents of regional change—the state, the local state and agent, and the foreign investor—is necessary for a fuller understanding of regional inequality in Jiangsu. During Mao's era, regional inequalities in Jiangsu increased. Although some efforts were made to develop backward counties, cities recorded faster growth than rural areas (represented by counties), due largely to Mao's industrialization policy, which favored the development of cities. In addition, local initiatives and decentralization policies facilitated the growth of Sunan. As a result, interregional, intercounty, and rural inequalities increased rather than declined.

During the reform period, central and local government policies in general favored the development of Sunan, especially Sunan's counties, while Subei was less favored in state policy. Sunan also has more favorable local development conditions, including local states, development experiences, and industrial structure. During the 1990s, intense reforms for decentralization, marketization, and globalization created favorable conditions for the development of non-state enterprises. The richest counties in Sunan, such as Wuxi, Wujiang, and Kunshan, registered much faster economic growth. Meanwhile, foreign investment also increased dramatically in Sunan. These factors have contributed to the rise of interregional and rural inequalities and the emergence of a group of counties in Sunan. The cities, that had the highest GVIAO per capita and grew faster than counties under Mao recorded slow growth. Strict central control, failures of reforming SOEs, and overburdened urban infrastructure have contributed to the slower growth of the cities, leading to the decline of intercounty inequality. An analysis of Kunshan has shown that regional development is a process incorporating global, state, and local forces, rather than a single factor. It is evident that economic reforms in China have produced a complex picture of regional inequality at the subprovincial level. While patterns of regional inequality may differ among the provinces, the findings of the functioning of multiple mechanisms and the significance of geographic scales in Jiangsu should help our understanding of regional inequality in China.

The intensification of the Sunan–Subei divide and rising regional conflicts have concerned the provincial government of Jiangsu.[11] Subei has argued forcefully that favorable reform and open door policies be extended to Subei and regional bias towards Sunan reduced. Corresponding to the Ninth FYP of China, which stressed the significance of the reduction of poverty and regional inequality, the provincial government of Jiangsu has also implemented its Ninth FYP which intends to accelerate the growth of Subei (People's Government of Jiangsu Province 1997). Indeed, since 1995, efforts have been made by the provincial government to support the development of poorer regions, including personnel support, fiscal subsidies, low-interest loans, and encouragement of interregional cooperation (Chu *et al.* 1997; Song *et al.* 1998). Considerable capital has been invested in infrastructure, agriculture, education, and industrial development. These efforts have improved the economies of poorer regions and have curbed the intensification of the Sunan–Subei gap, although the gap is not declining.

7 Conclusion

China since the late 1970s has implemented economic reforms and has experienced profound changes that require a careful assessment. In this research I have examined patterns and mechanisms of regional development and inequality in China, especially during the reform period. I have argued that three sets of mechanisms—the central state, local states and local conditions, and global forces—are particularly important to understanding the changing patterns of regional development in China. In this chapter, I summarize the major findings of this research and discuss its theoretical and policy implications.

7.1 Major findings

After the founding of the PRC in 1949, the Chinese government established a centralized planning and management system. Industrialization, national defense, and spatial equality were emphasized during the 1950s, and the conditions of the coastal industrial bases and some interior cities were improved. However, development policies under Mao were inconsistent and influenced by changing international and domestic conditions. While Mao stressed industrialization and national defense, decentralization policy in the late 1950s and early 1970s provided some decision-making power to local governments. During Mao's era, while the central state played a dominant role in economic and regional development, the role of local states and the international climate cannot be ignored.

When the PRC was established, Mao was troubled by a poor economy with a large coastal–interior disparity. During Mao's era, while some efforts were made to develop China's interior provinces, both interregional and interprovincial inequalities persisted. The level of interprovincial inequality in the mid-1970s (right before the launch of economic reforms) was even higher than that in the 1950s. Especially during the early 1970s, interregional and interprovincial inequalities even increased. These findings contrast sharply with the widely held notion that Mao's China had become more equal in regional development.

The persistence of regional inequality under Mao was the result of four major factors. First, China is a nation with diverse development conditions and a historical legacy of regional imbalance. As shown in many countries, such disparities are difficult to overcome in a short period of time. Secondly, considerations for industrialization and national defense were more important than equality in resource allocation. Mao's industrialization policy improved the economies of traditional industrial bases (e.g., Shanghai and Liaoning), which contributed greatly to the rise of interprovincial inequality under Mao, while interior-based defense spending contributed little to interior development. Thirdly, local forces and Mao's decentralization policy also contributed to the persistence of the coastal–interior inequality. With better development conditions, the coastal provinces captured the opportunities provided by decentralization policy to develop local economies. Lastly, political struggle, social unrest, poor planning, and economic mismanagement also contributed to the persistence of regional inequality. In short, uneven distribution of resources, consideration for industrialization and national defense, decentralization, and problems with orthodox socialism made the task of equalizing productive forces across regions difficult. Some of these factors can also be observed in the former Soviet Union, such as the emphasis on urban-centered industrialization, national defense, and political unrest, which contributed greatly to the persistence of income and regional inequalities in the former socialist countries in general.

Economic reforms have brought dramatic changes to the role of the state, market, and external forces. Although the central state remains important, local and global forces have emerged as critical currents underlying China's regional development. While the state is still playing an important role, the control and capacity of the central state have declined. State policies, including fiscal and investment reforms, the coastal development strategy, and the open door policy, have stimulated the growth of the coastal region. Traditional industrial bases and many poor interior provinces, however, have been tightly controlled by the central government and have received less favorable state support. Moreover, local states and agents have emerged as another important force driving economic and regional development in China.

Decentralization and marketization have provided considerable decision-making powers to local governments that have actively participated in the process of economic reforms and have initiated numerous policies for local economic development. Coastal provinces, with active local states and favorable local development conditions, are moving ahead of interior provinces in economic growth. Lastly, globalization and open door policies have contributed to the dramatic increase of foreign investment in China, especially in the coastal provinces, while the interior provinces lack foreign capital. Therefore, foreign investment, state policy, and local states and development conditions have all played important roles in uneven regional development in China. Consequently, the gap between the coastal region and the interior region has risen.

During the reform period, interprovincial inequality declined. This is not the result of any intention of the Chinese government to reduce regional inequality; rather, it is related to the slower growth of the traditionally richest provinces of China (i.e., Shanghai, Beijing, Tianjin, and Liaoning) and the rapid growth of the coastal provinces of Guangdong, Jiangsu, Zhejiang, Shandong, and Fujian. The growth rates of the traditional industrial centers and leading provinces were among the lowest. As was explained previously, the slow growth was caused mainly by the tight control and high taxation by the central state, the failure of reforming SOEs, and the deterioration of urban infrastructure in these provinces. As a result, the gap between the richest province and the poorest province has narrowed since the late 1970s.

On the other hand, the coastal provinces of Guangdong, Jiangsu, Zhejiang, Fujian, and Shandong have grown dramatically during the reform period. These provinces have more non-state enterprises and enjoy better local resource endowments. Local and provincial governments in these provinces have initiated numerous local development policies. Economic reforms have provided them with favorable policies in finance, investment, sectoral adjustment, and foreign investment and trade. With favorable local policies and investment conditions, these provinces have been favored by foreign investors. The rapid growth of the five coastal provinces has contributed to the faster growth of the coastal region as a whole, leading to the increase of the coastal–interior gap. But it also caused the relative decline of the four traditionally leading provinces (Shanghai, Beijing, Tianjin, and Liaoning), contributing to the decrease of interprovincial inequality. Without the consideration of the three municipalities (Shanghai, Beijing, and Tianjin), interprovincial inequality* in China increased slightly.

I have further examined intraprovincial inequality through a case study of Jiangsu. I have found that Mao's industrialization policy favoring cities resulted in the rapid growth of the cities in Jiangsu. Meanwhile, local initiatives and policies of decentralization and rural industrialization to a certain extent improved the conditions of Sunan. Therefore, during Mao's era, both state policy and local forces influenced regional development, while the global context cannot be ignored. The growth of cities and Sunan led to the rise of interregional, intercounty, and rural inequalities in Jiangsu under Mao, which is consistent with the rise of interregional and interprovincial inequalities in China.

During the reform period, the central state, local states and development conditions, and foreign investors have all favored the development of Sunan, while Subei has lagged far behind. TVEs, managed by local officials and based on local resources, have blossomed in Sunan. Sunan has also dominated foreign investment and trade. As a result, interregional inequality across Sunan, Suzhong, and Subei and rural intercounty inequality across counties increased. Overall intercounty inequality across all county-level units, however, has declined. This is related directly to the relative declining status of cities that have grown slower due to strict state control and problems with SOEs, while counties in Sunan have grown rapidly. Favorable macro policies, active local states, favorable local development conditions, and the infusion of foreign investment have led to the rapid growth of non-state sectors and the rise of a group of counties in Sunan. The rapid growth of counties in Sunan has also led to the rise of rural intercounty inequality.

7.2 Theoretical implications

As I have shown in Chapter 1, dominant theories of regional development are largely based on the assumptions of free market economies and the experiences of developed countries. Theories of regional inequality remain less sensitive to transition economies like China, which are undergoing dramatic restructuring and where factors of regional development differ considerable from Western countries. Moreover, the significant transformation of the global economy has made traditional theories less effective in explanation.

A central argument of this book is that in order to understand the change and complexity of regional inequality, we need to examine three major forces—the state, local states and agents, and globalization. China has been experiencing dramatic changes since the launch of economic reform in 1978, which emphasizes decentralization, marketization, and globalization. Consequently, global and local forces have emerged as major forces shaping China's regional inequality, while the state still plays an important role in economic development. Orthodox theories of regional inequality either totally ignore the role of the state or consider that the state merely plays a marginal role in regional development. In socialist countries, however, the state traditionally played a dominant role in regional development, and the Chinese state is still playing an important role in economic and regional development. Moreover, the state is not static, but shifting with the changing international and domestic conditions. Orthodox theories of regional development also tend to de-emphasize the role of foreign investment. In China, however, foreign investment has grown rapidly and has emerged as another major force underlying uneven regional development.

On the other hand, research on China tends to overemphasize the role of the central state in accounting for uneven regional development. While I acknowledge the critical role played by the central state, I have argued, in this book, that local states and factors are another important force underlying China's regional development. During Mao's era, resource-rich provinces, such as Liaoning, benefited greatly from their resource endowments and the allocation of state investment. During the reform period, local governments in coastal provinces, such as in Jiangsu and Zhejiang, have actively initiated numerous policies to stimulate local economic growth, while interior provinces have received less favorable state policy and have been less effective in stimulating local economic development. In particular, those provinces dominated by SOEs, such as China's traditional industrial bases—for example, Liaoning—have lagged behind. The implication is that both the central state and local states are playing important roles in economic and regional development, rather than either the central state or the local state.

In addition, globalization and China's reforms have stimulated the rapid growth of foreign investment in China, which has emerged as another major force underlying China's regional development. The coastal provinces have captured the lion's share of foreign investment in China, partially as a result of a favorable open door policy, local initiatives, and local development conditions. Therefore, the central state, local states and agents, and foreign investment have all contributed to uneven regional development in China. A synthesis of global, state, and local forces better explains the changing patterns of regional inequality. The relationship between regional inequality and its causing factors is more complicated than what theories of regional inequality have often assumed. A notion of multi-mechanisms, rather than focusing on a single factor, should provide a fuller interpretation of regional inequality.

This research has shown that regions that have received preferential state policies, attracted more foreign investment, and provided favorable local conditions tend to grow faster than others. In China, traditional industrial bases, with a large state sector and being favored by Mao's industrialization policy, have been slower in adapting to the emerging market mechanisms. The fortune of regions is changing with the changing domestic and international conditions, and the wealth of regions is related to local development conditions. As I have revealed in this book, a complex mosaic of regional development has emerged in China. Previous studies on China tend to oversimplify patterns of regional development and the change of regional inequality.

Regional inequalities in China and in Jiangsu for the last four and a half decades show no clear divergent, convergent, or inverted-U patterns. In China, while interregional inequality rose over time, interprovincial inequality increased under Mao but decreased in the post-Mao period, produced by the shifting development conditions and changing fortune of the regions. In Jiangsu, interregional, intercounty, and rural inequalities all increased under Mao. During the reform period, however, while interregional and rural inequalities have risen further, intercounty inequality has declined. Geographic scales are important to the study of regional inequality, and the notion of convergence, divergence, or inverted-U pattern is oversimplified in describing the trend of regional inequality. In summary, my analysis of China shows that regional inequality is shaped by multi-mechanisms, related to the fortune of regions, and is specific to geographic scale. The rigidity of assumptions underlying convergence or divergence schools limits their application in understanding the empirics of regional inequality.

7.3 Policy implications

The intensification of China's coastal–interior inequality, particularly in the 1990s, has been much resented by the poor provinces. Some argue that state policy favoring the coastal region is fundamental to the rise of the inequality. Regions have bargained intensely with the central government over the level of central control and regional distribution of resources. Considering that they are being unfairly treated by the central government, many interior governments have used their increasing administrative powers to protect their resources and markets, by distorting price signals and creating unfair competition to limit market penetration of industrial products produced by other regions. Local protectionism and 'investment hunger' have prompted regions to duplicate investment projects with little consideration for economies of scale and agglomeration.

Since the mid-1990s, the central government has been increasingly concerned with the rise of interregional inequality, the weak trickle-down effects, and the increase of regional conflicts. This awareness has partly arisen from the lobbies of the interior provinces for extending preferential policies to the interior and the intellectual critique on unequal interregional competition. Consequently, many interior areas have been opened up for foreign investment and trade, and the extent of policy tilting toward the coastal region has been reduced. The Ninth FYP (1996–2000) acknowledges the importance of interior development and promotes regional cooperation to avoid excessive regional income gaps that may further intensify social problems and interregional conflicts.

Indeed, the central government has never totally ignored the issue of regional inequality. Even during the 1980s when development policy was biased towards the coastal region, some measures were taken to reduce the coastal–interior divide. Special funds, preferential bank loans, and tax exemption policies have been continuously granted to the interior. The government encouraged the coastal provinces to make compensation agreements on energy and raw materials with the interior provinces. Economic reforms have been gradually extended to the interior areas. In the early 1990s, the State Council had taken more aggressive measures to help the interior to develop TVEs. With the increasing revenue as a result of income growth and tax reforms, state investment in the interior region has increased since the mid-1990s, especially for the improvement of education, infrastructure, energy, and raw materials. The central government has urged coastal enterprises and foreign investors to provide capital and management skills to improve the conditions of the interior provinces. The interior has also experienced some growth in exports and foreign investment.

However, the macro context for China's regional development has changed rapidly. While the state remains important, the control of the central state over the Chinese economy has declined. The central government no longer monopolizes the resource allocation process and has become less capable of commiting a large percentage of investment for interior development. The central state could never ignore the development of the coastal provinces, which are the major tax contributors to the state coffers and which can participate more effectively in the international division of labor. This is evidenced by the vigorous development of Shanghai's Pudong district and the drive to build Shanghai into a global city. Meanwhile, as a result of decentralization, marketization, and globalization, local and global forces have emerged as important factors underlying economic and regional development. Coastal areas have more favorable local states and development conditions, which have become more important than ever in regional development. Foreign investment always favors some regions over others and tends to concentrate in China's coastal provinces and well-developed regions.

While an outline of China's future regional policies is beyond the task of this book, domestic restructuring and global changes require new strategies of regional development. While some scholars have argued for increasing fiscal transfers to reduce regional inequality, regional policy no longer emphasizes redistribution. Although the central government has encouraged the development of TVEs in the interior provinces, evidence suggests little progress has been made. Regional development no longer relies on the policy of the central government, but has to consider local efforts and global changes. Non-profit organizations, non-state enterprises, and individuals should engage more actively in regional development, although the government is still important in improving infrastructure and human capital. Foreign investment and trade are also important in the development of poorer regions. It is expected that China's regions will continue to be fragmented, and regional inequality will remain an important issue facing researchers and the Chinese government.

Notes

1 Introduction

1 Region is often defined as a geographic area of the world, and can refer to place or locality. Regions, as well as localities, have also been conceptualized as processes and agents operating locally (see Note 4 for locality). Both concepts of regions are used in this book. For the purpose of measuring regional inequality, administrative units are often used, mainly for data and policy reasons. See Richardson (1978), Markusen (1987), and Johnston *et al.* (1994) for further discussion on the concepts of regions.

2 Examples of regional 'resource wars' include the 'wool wars' and 'silk cocoon wars' in which resource-rich regions used their powers, even the local branches of the People's Liberation Army, to block the outflow of their resources. For more information, see Wedeman (1993) and Cannon and Zhang (1996).

3 I use the term the 'central state' or 'state' to refer to the central government. The literature on the Chinese state is extensive, e.g., Shue (1988); Oi (1989); Goodman and Hooper (1994); Walder (1995a, 1995b).

4 Locality, in lay terms, simply refers to region or place, and has been used as policy for local states, local labor markets, spatial variation, or anything to do with the local (Johnston *et al.* 1994). In this book, I use locality to refer to local states and local development conditions.

5 I am fully aware of the importance of consumption and social indicators. However, as this book focuses on regional economic/income inequality, social indicators, such as health care, are beyond this research. Elsewhere, I have reported my findings on spatial inequality in heath care facilities in China (e.g., Wei and Dutt 1992).

6 National income (NI), or net material product (NMP), refers to the newly created value in a given period by the workers engaged in material production. It is the same as gross value of output less the values of the materials consumed during the process of production.

7 Gross value of industrial and agricultural output (GVIAO) includes gross value of industrial output (GVIO) and gross value of agricultural output (GVAO). GVIO refers to the total value of the finished and semi-finished industrial products by enterprises in the industrial, mining, and energy sectors. GVAO refers to the total value of production generated by farming, forestry, animal husbandry, sideline activities, and village industries.

8 Data involving temporal comparison are in comparable prices (*kebi jiage*). The State Statistical Bureau of China releases income and output data at comparable prices in index form. These indexes can be used to calculate real NI, real GDP, and real GVIAO, as Tsui (1991) and Chen and Fleisher (1996) did. The base year

for the real NI and real GDP is 1952, and the base year for the real GVIAO in Jiangsu is 1950.

9 Fixed asset investment (FAI) includes capital construction investment (CCI) and investment for project renovation and transformation. Capital construction investment in state-owned enterprises (CCISOE) has been the major vehicle for the Chinese government to implement its development policies. It refers to the allocation of funds by the government to construct or enlarge state-owned enterprises and their related facilities.

10 For a discussion on the limitations of labor mobility within China, see Jian *et al.* (1996). Elsewhere, I have also reported my work on migration and urban development in China (e.g., Wei 1989, 1994, 1997b).

11 Two other less often used regional classifications include the north and south regions and aggregating China's provinces into 6–8 economic regions; see Wei (1993), Hu *et al.* (1997), and Hu (1998) for regionalization.

12 To maintain consistency, the data for Hainan, which was separated from Guangdong province in 1988, will sometimes be combined with those for Guangdong and will be specified. Chongqing will be combined with Sichuan.

13 A number of indexes are available for the analysis of regional inequality, including the coefficient of variation (CV), the Gini coefficient, and Theil's entropy index (Coulter 1989). I have calculated the above three indexes and found that the broad patterns are similar. While the Gini coefficient is often used for measuring income inequality, CV is the most often-used index to measure regional inequality. The equation for CV is as follows:

$$CV = \frac{\sqrt{\sum(X_i - \bar{X})^2/N}}{\bar{X}}$$

where N = number of regions, X_i = observation of the ith region, and $\bar{X}$ = arithmetic mean for variable X_i. The higher the CV, the greater the inequality.

14 The location quotient (LQ) for national income is calculated as follows:

$$LQ = \frac{X_i/\sum X_i}{Y_i/\sum Y_i}$$

where X_i and $\sum X_i$ are the regional and total amounts of national income (NI), and Y_i and $\sum Y_i$ are the regional and total population. Therefore, the LQ is also equal to the regional per capita NI divided by the average per capita NI. An LQ over 100 (or 1) shows that a region produces a higher than the average per capita NI, while less than 100 (or 1) indicates the opposite.

15 Geographic information system (GIS) is defined as the process of and sequence of components for acquiring, processing, storing, and managing spatial data. It is revolutionizing geography, and is one of the most important developments in information technology.

2 Patterns of regional development

1 The significance of colonial forces in creating regional imbalance, however, should not be exaggerated, as the disparity between China proper and extra-ecumenical China existed before the penetration of colonial forces, and physical conditions also provide some foundations for the imbalance.

2 During Mao's era, China's Five-Year Plans (FYPs), except the First FYP, were poorly designed, and implementation was severely interrupted by political struggle and social unrest. For more information on China's FYPs, see Larsen (1992).

3 The Soviet-model of industrialization emphasizes rapid industrial growth centered on the development of heavy industries, particularly the iron and steel, oil, and heavy chemical industries.
4 The Soviet principles of industrial allocation include: (1) production should be sited in proximity to major raw material and energy sources; (2) production of manufacturing and services should be near consumption centres; (3) economic regions should specialize in the production of the most efficient goods; (4) economic activity should be evenly distributed throughout the country and reduce spatial inequalities; (5) industrial allocation must be consistent with the need of national defense, and so on. See Schiffer (1989).
5 Other than specified, all data are from the SSB's annual statistical yearbooks.
6 The cancellation of numerous capital construction projects, the closing of thousands of factories, and the curtailment of many others left millions of urban workers unemployed, and many were sent to rural areas, due also to the shortage of food and services.
7 Most of the problems described here are typical of socialist countries. For a more detailed analysis of the problems of the socialist system, see Kornai (1992).
8 Tibet was excluded due to the lack of data. Hainan, which was separated from Guangdong in 1988, was combined with Guangdong for temporal comparison.
9 The description of regional inequality is largely based on the change of CVs.
10 The smaller value of CVs for regions is related to the small number of sample size (three regions). China's interprovincial inequality was among the highest in the world. The CV for the United States, for example, was 0.216 in 1950, 0.124 in 1978, and 0.161 in 1991 respectively (Crown and Wheat 1995). In the Soviet Union, the CV for the republics based on national income was 0.28 in 1970 (Bahry and Nechemias 1981).
11 CVs are sensitive to extreme values. So the changing status of the leading provinces is particularly important to the change of CVs.
12 This number is not small, as the unit of measurement is the three regions.
13 The 'coastal' group is for the analysis of this section only, and should not be confused with the coastal region.

3 Fiscal systems, investment, and regions

1 Readers should be careful with Mao's slogans, as they might be very much different from what was really implemented.
2 For further information on vertical (*tiao*) and horizontal (*kuai*) government organizations in China, see Lyons (1987).
3 These problems are typical to Soviet-type socialist economies, see Kornai (1992).
4 The extrabudgetary fund was subject to less stringent direct control from the state but was still under the scope of state planning. The policy towards extrabudgetary funds, however, has been shifting constantly.
5 This approach has been used by many scholars, such as Ferdinand (1989). But with the increasing non-balancing financing, it has become less effective in indicating the level of interregional budget transfer.
6 In principle, the central government requires the provinces to balance their budget. This target, however, was often not met, partly due to lower revenue or overspending by local governments. The central government itself often cannot balance its budget mainly due to overspending.
7 The level of price distortion is extremely difficult to measure, as historical data are lacking and scholars disagree over the 'fair' price structure. Reforms have raised the prices of agricultural products and raw materials, and the problem of price distortion has become less serious. For some work related to this issue, see Chai (1997) and Yang (1997).

8 The rate of interregional national income transfer = (national income used − national income)/national income × 100.
9 The significance of venture capital in regional development may have been exaggerated, since little empirical work has dealt with total investment and other forms of investment.
10 These problems are typical to Soviet-type command economies (Kornai 1992).
11 Prior to 1978, fixed and working capital was financed by the state budget, and enterprises had to remit all profits to the state (zero cost of capital). Reforms gradually reduced budgetary grants and channeled funds through state-controlled banks which lent funds with certain interests (nonzero cost of capital).
12 Key-point (construction) projects refer to large-scale, centrally controlled investment projects, which reflect the policies of the central government on investment and economic growth, such as the 156 projects during the First Five-Year Plan (1953–57). These projects are usually in sectors with strategic significance, such as energy, transportation, and technology. However, total investment, sectoral allocation, and geographic location of the key-point projects change from year to year.
13 Each province was also divided into smaller 'third, second, and first fronts.' Strategic or defense-oriented projects were often built in the small third front (*xiao san xian*).
14 An investment cycle is considered as a driven force underlying China's economic cycle; see Imai (1994).
15 These categories of investment are based on the classification of the State Statistical Bureau of China. Enterprise funds are also sometimes translated into self-raised funds (*zichou zijin*).

4 Ownership, foreign investment, and beyond

1 Chinese official statistics classify the industry into four groups by ownership: state-owned enterprises (SOEs), collective-owned enterprises (COEs), individual-owned (private) enterprise (IOEs), and other enterprises (mainly foreign-invested enterprises (FIEs) and shareholding enterprises (SHEs)) (SSB 1996; Jefferson and Singh 1999). According to the State Statistical Bureau (SSB 1996, p. 450), SOEs refer to enterprises where means of production or income are owned by the state. The administration of SOEs, however, is more complicated as some are administered by the central government, and an increasing number of SOEs are administered by local governments (provincial, prefecture/municipal, and county governments).
2 However, we should be careful as policies under Mao were changing constantly and he tended to be biased towards certain sectors during certain periods.
3 Chinese statistics published by the State Statistical Bureau did not show private enterprises existing during Mao's era. In Wenzhou region, however, as witnessed by the author, the underground capitalist economy was never totally eliminated. Also see Liu (1992).
4 The non-state sector refers to enterprises other than SOEs, mainly including COEs, IOEs and FIEs. According to SSB (1996, p. 450), COEs, including urban COEs and rural COEs, refer to enterprises where the means of production are owned collectively, rather than by the state. Rural COEs (often called township and village enterprises (TVEs)) are owned by township and village citizens and largely controlled by township and village governments. Enterprises of other ownership mainly consist of foreign-invested enterprise (FIEs) which are also called 'three capital enterprises' (*sanzi qiye*) denoting (1) joint ventures, (2) contractual joint ventures, and (3) wholly foreign-owned enterprises.

5 The tax for profit system (*li gai shui*) was introduced in 1984. It substitutes tax payments for profit remittances in an attempt to revitalize SOEs. This system reclassifies profits into a set of functionally differentiated revenue streams, mainly taxes; and enterprises retain after-tax profit, rather than handing all profit to the government.
6 Proportions of SOEs in fixed assets are even higher than in output, as much of the earlier fixed investment went to SOEs; see Jin and Chen (1996).
7 This is one of the major differences between China and the Soviet Union in their experiences with economic reforms, and contributes greatly to the success of China's economic reforms.
8 There is a large body of literature on special economic zones, e.g., Jao and Leung (1986), Crane (1990), and Park (1997).
9 For more information on regulations for foreign investment, see Grub and Lin (1991), Wang (1997), and Luo (1998).
10 Investment from Hong Kong is treated as foreign as its behavior is still very different from China's domestic investment, although Hong Kong was returned to China in 1997.
11 Earlier figures of investment from Hong Kong as released by the SSB included a small volume from Macao.
12 By 1992, China had established 8700 development districts. In 1992 alone, developing the open districts occupied 1.5 million hectares of land, which was similar in size to 46 Shenzhen special economic zones, or half of Hainan province (H. Wei 1995).
13 As my model focuses on capital investment, the effects of local resource endowments, such as education and transportation, are not included. Locational determinants of regional growth are well documented in economic and geographical literature. For more information, see Armstrong and Taylor (1985), Malecki (1991), and Crown and Wheat (1995).
14 The introduction of a coastal dummy captures China's coastal–interior disparity and improves regression analysis. However, the implications of this variable remain debatable. For example, both Gong (1995) and Jian *et al.* (1996) used a coastal dummy to represent China's open door policy.
15 Other independent variables may include education, transportation and foreign trade.

5 Patterns of regional development in Jiangsu

1 The usage of these three terms—interregional inequality, (overall) intercounty inequality and rural (intercounty) inequality—is for this book only. There is no consensus regarding proper terms used to describe regional inequalities within provinces. For some scholars, intercounty inequality refers to inequality across all county-level administrative units, while for others, intercounty inequality refers to inequality across counties (including county-level cities). Strictly speaking, regional inequality across counties and county-level cities cannot be named as rural intercounty inequality. This is because a county-level city is a unique county-level administrative unit (if compared with counties or cities in the United States), and its economy is dominated by urban-oriented industries and services, rather than agriculture or traditional rural industries in rural areas.
2 Readers have to pay special attention to the different usage of the term intercounty inequality. Rozelle (1994), for example, called inequality across counties (including county-level cities), intercounty inequality. Fan (1995b) and Wei and Fan (1997) refer to inequality across all basic administrative units (counties, county-level cities, and city districts) as intercounty inequality.

3 Jiangsu is traditionally divided into Sunan and Subei via the Yangtze River. However, as Sunan often refers to *su-xi-chang* (Suzhou, Wuxi, and Changzhou), many scholars have adopted a three-division approach, which divides Jiangsu into Sunan, Suzhong, and Subei. See Fig. 5.1 for the division.

4 It is difficult, however, to know exactly how much resources the provincial government has allocated to the development of Subei.

5 Conflicts arising from the 'city-leading-county' system have generated considerable governmental concerns. But few Western scholars have studied this system. For some limited information, see Ma and Noble (1986) and Wei (1994).

6 City districts (*shiqu*) of municipalities are previous prefecture seats, which are under the direct jurisdiction of the province and are often referred to as prefecture-level cities (*di ji shi*). In 1995, Jiangsu was divided into 11 municipalities, which include 11 municipal-level city districts, 32 county-level cities, and 32 counties. Two new municipalities of Taizhou (separated from Yangzhou) and Suqian (separated from Huaiyin) were established in 1996.

7 See Chapter 5 notes 1 and 2 regarding terms given to different levels of regional inequality.

8 GVIAO has traditionally been used as the most important indicator of the Chinese economy at the county level. For details regarding the effectiveness of this indicator, see Rozelle (1994).

9 For example, in 1931, the flooding in the Lixiahe area led to 77,000 deaths and the out-migration of 1.6 million people (Editorial Committee 1989).

10 For example, the Gengche model is well known for its economic development based on both rural agricultural and industrial resources. For models of regional development in China, especially the Wenzhou model, see Wei (1989), Nolan and Dong (1990), and Zhou and Zhang (1994).

11 A comparison of Suzhou with Wenzhou can find that Suzhou develops ahead of Wenzhou particularly in terms of social and development indicators, see Zhou and Zhou (1997).

12 This number is considered as very large, given that interregional inequality is calculated based on three regions.

13 ** indicates significance at the 1 percent level.

14 Some scholars call the success of Yangzhou (in Suzhong) the Yangzhou model or the 'Yangzhou phenomenon' (*Yangzhou xianxiang*).

6 Mechanisms of regional development in Jiangsu

1 This contrasts sharply with the local states in most of the developed countries that act as instruments to provide public goods and to promote public–private relations.

2 'One contract' refers to the economic contract system; 'three reforms' introduced the contract system to official appointment, labor recruitment, and workers' wages.

3 Naughton (1995b), for example, argues that the bulk of rural industrialization might best be thought of as 'suburban industrialization.'

4 The provinces with higher per capita revenue in 1978 include Shanghai, Beijing, Tianjin, Liaoning, Heilongjiang, and Gansu. All of them are China's major industrial centers dominated by state-owned enterprises and heavily invested in by the state during Mao's era.

5 In 1996, Guangdong's extrabudgetary revenue was 24.8 billion yuan, lower than Jiangsu, but its expenditure was higher than Jiangsu (24.9 billion yuan) (MOF 1997). Several other coastal provinces also registered high extrabudgetary funds, such as Shandong and Zhejiang.

6 Suzhou also dramatically adjusted the leadership of TVEs by appointing younger, better-educated, more skilled, and more entrepreneurial leaders.

7 China's national urban policy intends to control the size of large cities and to promote the development of small cities; see Wei (1994) and Zhao and Zhang (1995).

8 For economic restructuring in Asian NICs and the outflow of capital from these countries, see, for example, Yeung (1994).

9 Lianyungang was opened up to foreign investment and trade in 1984, as one of the 14 open coastal cities in China. The construction of its development district began in 1985, with a planning area of 45 km^2. However, it has absorbed much less FDI than the major development districts in Sunan. In 1996, it attracted US$40 million FDI, much less than development districts in Suzhou (US$410 million), Kunshan (US$380 million), and Wuxi (US$255 million) (People's Government of Jiangsu Province 1997).

10 For definitions of interregional, intercounty, and rural inequalities in Jiangsu, see Notes 5.1 and 5.2.

11 Interview with local officials in July 1998.

Bibliography

Agarwala, R. 1992. *China: Reforming Intergovernmental Fiscal Relations.* Discussion Papers 178. Washington DC: China and Mongolia Department, World Bank.

Agnew, J. and Corbridge, S. 1995. *Mastering Space.* New York: Routledge.

Aguignier, P. 1988. Regional disparities since 1978. In Feuchtwang, S., Hussain, A. and Pairault, T. (eds), *Transforming China's Economy in the Eighties*, Vol. 2, 93–106. Boulder, CO: Westview Press.

Alonso, W. 1980. Five bell shapes in development. *Papers, Regional Science Association* 45: 5–16.

Amin, A. and Thrift, N. (eds) 1994. *Globalization, Institutions, and Regional Development in Europe.* Oxford: Oxford University Press.

Amos, O.M., Jr. 1990. Growth pole cycles. *Review of Regional Studies* 20(1): 37–48.

Appelbaum, R.P. and Henderson, J. (eds) 1992. *States and Development in the Asian Pacific Rim.* Newbury Park, CA: Sage.

Armstrong, H. and Taylor, J. 1985. *Regional Economics and Policy.* Oxford: Philip Allan.

Arthur, W.B. 1994. *Increasing Returns and Path Dependence in the Economy.* Ann Arbor: University of Michigan Press.

Auty, R. 1992. Industrial policy reform in China. *Transactions, Institute of British Geographers, N.S.* 17: 481–94.

Bahl, R. and Wallich, C. 1992. *Intergovernmental Fiscal Relations in China.* WPS 863. Washington, DC: Country Economics Department, World Bank.

Bahry, D. and Nechemias, C. 1981. Half full or half empty? *Slavic Review* 40(3): 366–83.

Barro, R.J. and Sala-i-Martin, X. 1991. Convergence across states and regions. *Brookings Papers of Economic Activity* 1: 107–82.

Barro, R.J. and Sala-i-Martin, X. 1992. Convergence. *Journal of Political Economy* 100: 223–51.

Barro, R.J. and Sala-i-Martin, X. 1995. *Economic Growth.* New York: McGraw-Hill.

Baum, R. 1994. *Burying Mao.* Princeton: Princeton University Press.

Beresford, M. and McFarlane, B. 1995. Regional inequality and regionalism in Vietnam and China. *Journal of Contemporary Asia* 25(1): 50–72.

Bird, R. and Wallich, C. 1994. Local finance and economic reform in Eastern Europe. *Environment and Planning C: Government and Policy* 12: 263–76.

Bluestone, B. and Harrison, B. 1982. *The Deindustrialization of America.* New York: Basic Books.

Bo, Y. 1993. *Ruogang zhongda juece yu shijian de huigu* (*Reflections of Some Important Decisions and Events*). Beijing: School of the Communist Party of China Press.

Borts, G.H. and Stein, J.L. 1964. *Economic Growth in a Free Market*. New York: Columbia University Press.

Bramall, C. 1993. *In Praise of Maoist Economic Planning*. Oxford: Oxford University Press.

Bramall, C. and Jones, M.E. 1993. Rural income inequality in China since 1978. *Journal of Peasant Studies* 21(1): 41–70.

Cai, L., Zhang, G. and Xu, Q. (eds) 1995. *Shanghai*. Shanghai: Shanghai People's Press.

Cannon, T. 1990. Regions: spatial inequality and regional policy. In Cannon, T. and Jenkins, A. (eds), *The Geography of Contemporary China*, 28–60. New York: Routledge.

Cannon, T. and Jenkins, A. (eds) 1990. *The Geography of Contemporary China*. New York: Routledge.

Cannon, T. and Zhang, L. 1996. Inter-region tension and China's reforms. In Cook, I.G., Doel, M.A. and Li, R. (eds), *Fragmented Asia*, 75–101. Aldershot: Avebury.

Chai, J. 1996. Divergent development and regional income gap in China. *Journal of Contemporary Asia* 26(1): 46–59.

Chai, J. 1997. *China: Transition to a Market Economy*. New York: Oxford University Press.

Chatterjee, L. 1989. Third World cities. In Peet, R. and Thrift, N. (eds), *New Models in Geography*, 127–46. London: Unwin Hyman.

Chen, J. and Fleisher, B.M. 1996. Regional income inequality and economic growth in China. *Journal of Comparative Economics* 22(2): 141–64.

Chen J. *et al.* (eds) 1995. *1995 nian zhongguo guomin jingji he shehui fazhan baogao* (*China National Economy and Social Development Report, 1995*). Beijing: China Planning Press.

Cheung, P.T.Y. 1996. The political context of Shanghai's economic development. In Yeung, Y. and Sung, Y. (eds), *Shanghai*, 49–92. Hong Kong: Chinese University Press.

Chi, C. 1936. *Key Economic Areas in Chinese History*. London: Allen and Unwin.

Chu, D. *et al.* (eds) 1997. *Benxiang xiaokang yu xiandaihua de Jiangsu* (*Jiangsu Towards a Comfortable Life and Modernization*). Nanjing: Jiangsu People's Press.

Chung, J.H. 1994–95. Beijing confronting the provinces. *China Information* 9(2/3): 1–23.

Clark, C. 1983. Regional inequality in communist nations. In Nelson, D.N. (ed.), *Communism and the Politics of Inequalities*, 73–104. Lexington, MA: Lexington Books.

Clark, G.L., Gertler, M.S. and Whiteman, J. 1986. *Regional Dynamics*. Boston: Allen and Unwin.

Cole, J.P. 1987. Regional inequalities in the People's Republic of China. *Fijdschrift voor Economische en Sociale Geografie (Journal of Economic and Social Geography)* 78: 2011–13.

Coulter, P.B. 1989. *Measuring Inequality: A Methodological Handbook*. Boulder: Westview Press.

Crane, G.T. 1990. *The Political Economy of China's Special Economic Zones*. Armonk, NY: M.E. Sharpe.

Crown, W.H. and Wheat, L.F. 1995. State per capita income convergence since 1950. *Journal of Regional Science* 35(4): 527–52.

Csefalvay, Z. 1994. The regional differentiation of the Hungarian economy in transition. *GeoJournal* 32: 351–61.

Denny, D.L. 1991. Regional economic differences during the decade of reform. In Joint Economic Committee, Congress of the United States, *China's Economic Dilemmas in the 1990s*, 186–208. Washington, DC: US Government Printing Office.

Dicken, P. 1998. *Global Shift*. London: Paul Chapman.

Ding, S. 1996. An analysis of China's northeast and the formation of the 'northeast phenomena.' In Li, S., Tang, W.S., Jiang, L. and Zhou, S. (eds), *Zhongguo quyu jingji fazhan mianmian guan (Aspects of China's Regional Economic Development)*, 155–73. Taibei and Hong Kong: Center for Population Research of National Taiwan University and East-West Center of Hong Kong Batist University.

Donnithorne, A. 1972. China's cellular economy. *China Quarterly* 52: 605–19.

Donnithorne, A. 1976. Centralization and decentralization in China's fiscal management. *China Quarterly* 66: 328–54.

Du, W. and Gu, J. (eds) 1987. *Zhongguo renkou: Jiangsu fence (Population in China: Jiangsu)*. Beijing: China Financial Economy Press.

Dunning, J. 1993. *Multinational Enterprises and the Global Economy*. Reading: Addision Wesley.

Dwyer, D. 1993. China: the consequences of liberalization. In Dixon, C. and Drakakis-Smith, D. (eds), *Economic and Social Development in Pacific Asia*, 152–68. New York: Routledge.

Editorial Committee. 1989. *Dangdai zhongguo de Jiangsu (Contemporary China: Jiangsu)*. Beijing: China Social Science Press.

Edmonds, R.L. 1990. History. In Cannon, T. and Jenkins, A. (eds), *The Geography of Contemporary China*, 61—79. New York: Routledge.

Emmanuel, A. 1972. *Unequal Exchange*. New York: Monthly Review Press.

Engels, F. 1975. *Anti-Duhring*. Moscow: Progress Publishers.

Evans, P. 1979. *Dependent Development: the Alliance of Multinational, State, and Local Capital in Brazil*. Princeton, NJ: Princeton University Press.

Fan, C.C. 1995a. Of belts and ladders: state policy and uneven regional development in post-Mao China. *Annals of the Association of American Geographers* 85: 421–49.

Fan, C.C. 1995b. Spatial impacts of China's economic reforms in Jiangsu and Guangdong provinces. *Chinese Environment and Development* 6(1 & 2): 85–116.

Fan, C.C. 1997. Uneven development and beyond: regional development theory in post-Mao China. *International Journal of Urban and Regional Research* 21: 620–39.

Fang, S. 1992. The growing role of township enterprises in Mainland China's economy. *Issues and Studies* 28(2): 128–30.

Ferdinand, P. 1989. The economic and financial dimension. In Goodman, D.S.G. (ed.), *China's Regional Development*, 38–56. New York: Routledge.

Fields, G. and Schultz, T.P. 1980. Regional inequality and other sources of income variation in Colombia. *Economic Development and Cultural Change* 3: 447–67.

Florida, R. and Smith, D.F., Jr. 1993. Venture capital formation, investment, and regional industrialization. *Annals of the Association of American Geographers* 83(3): 434–51.

Forbes, D. and Thrift, N. 1987. *The Socialist Third World*, 53–97. New York: Basil Blackwell.

Frank, A.G. 1967. *Capitalism and Underdevelopment in Latin America.* New York: Monthly Review Press.

Freeman, C.P. 1998. China's reform challenge: the political economy of reforming northeast China. Paper presented at the 47th Annual Midwest Conference on Asian Affairs, University of Wisconsin-Milwaukee, September 25–27.

Friedman, E., Pickowicz, P. and Selden, M. 1991. *Chinese Village, Socialist State.* New Haven, Conn.: Yale University Press.

Friedmann, J. 1966. *Regional Development Policy.* Boston: MIT Press.

Friedmann, J. 1986. World city hypothesis. *Development and Change* 17: 69–83.

Fuchs, R.J. and Demko, G.J. 1979. Geographic inequality under socialism. *Annals of the Association of American Geographers* 69(2): 304–18.

Gertler, M.S. 1986. Regional dynamics of manufacturing and non-manufacturing investment in Canada. *Regional Studies* 20(6): 523–34.

Gilbert, A.G. and Goodman, D.E. 1976. Regional economic disparities and economic development. In Gilbert, A.G. (ed.), *Development Planning and Spatial Structure,* 113–41. New York: John Wiley & Sons.

Gong, H. 1995. Spatial patterns of foreign investment in China's cities, 1980–1989. *Urban Geography* 16(3): 198–209.

Goodman, D.S.G. 1989. Political perspectives. In Goodman, D.S.G. (ed.), *China's Regional Development,* 20–37. London: Routledge.

Goodman, D.S.G. (ed.) 1997. *China's Provinces in Reform.* London: Routledge.

Goodman, D.S.G. and Hooper, B. (eds) 1994. *China's Quiet Revolution.* New York: St. Martin's Press.

Goodman, D.S.G. and Segal, G. (eds) 1994. *China Deconstructs.* New York: Routledge.

Gore, C. 1984. *Regions in Question.* New York: Methuen.

Green, A.G. 1969. Regional inequality, structure change and economic growth in Canada 1890–1956. *Economic Development and Cultural Change* 17: 567–83.

Granovetter, M. 1985. Economic action and social structure: the problem of embeddedness. *American Journal of Sociology* 91(3): 481–510.

Gregory, D. and Urry, J. (eds) 1985. *Social Relations and Spatial Structures.* London: Macmillan.

Grub, P.D. and Lin, J.H. 1991. *Foreign Direct Investment in China.* New York: Quorum Books.

Gu, S. and Shen, L. (eds) 1989. *Jiangsu jinji shenian chonglun* (*Ten Years of Jiangsu's Economy*). Nanjing: Nanjing Press.

Guo, W. 1988. The transformation of China's regional policy. *Development Policy Review* 6: 29–50.

Gurtov, M. and Hwang, B.M. 1980. *China under Threat.* Baltimore: Johns Hopkins University Press.

Hansen, N. 1981. Development from above. In Stohr, W.B. and Taylor, D.R.F. (eds), *Development from Above or Below?* 15–38. New York: John Wiley & Sons.

Harvey, D. 1982. *The Limits to Capital.* Oxford: Basil Blackwell.

Hausner, J., Jessop, B. and Nielsen, K. (eds) 1995. *Strategic Choice and Path-Dependency in Post-Socialism.* Vermont: Edward Elgar Publishing.

Higgins, B. and Savoie, D.J. 1995. *Regional Development Theories & Their Applications.* New Brunswick: Transaction Publishers.

Hirschman, A.O. 1958. *The Strategy of Economic Development.* New Haven: Yale University Press.

Ho, S.P.S. 1994. *Rural China in Transition.* Oxford: Clarendon Press.

Hodder, R. 1994. State, collective and private industry in China's evolving economy. In Dwyer, D. (ed.), *China: The Next Decades,* 116–27. Essex: Longman.

Honing, E. 1992. *Creating Chinese Ethnicity.* New Haven: Yale University Press.

Howell, J. 1991. The impact of the open door policy on the Chinese state. In White, G. (ed.), *The Chinese State in the Era of Economic Reform,* 119–46. Armonk, NY: M.E. Sharpe.

Howell, J. 1993. *China Opens its Doors.* Colorado: Lynne Rienner Publishers.

Hsing, Y. 1998. *Making Capitalism in China.* New York: Oxford University Press.

Hsu, R.C. 1991. *Economic Theories in China, 1979–88.* Cambridge: Cambridge University Press.

Hu, A., Wang, S. and Kang, X. 1995. *Zhongguo diqu chayi baogao (Report on Regional Disparities in China).* Shenyang: Liaoning People's Press.

Hu, X. 1998. *Quyu yu Chengshi Yanjiu (Regional and Urban Studies).* Beijing: Science Press.

Hu, Z., Wang, E. and Han, M. 1997. *Zhongguo Quyu Jingji Uayi jiqi duice (China's Regional Economic Inequality and Strategies).* Beijing: Qinghua University Press.

Huang, Y. 1996. *Inflation and Investment Controls in China.* Cambridge: Cambridge University Press.

Huo, S. 1994. Regional inequality variations and central government policy, 1978–1988. In Hao, J. and Lin, Z. (eds), *Changing Central-Local Relations in China,* 182–206. Boulder, CO: Westview Press.

Imai, H. 1994. China's endogenous investment cycle. *Journal of Comparative Economics* 19: 188–216.

Jacobs, J.B. and Hong, L. 1994. Shanghai and the Lower Yangzi Valley. In Goodman, D.S.G. and Segal, G. (eds), *China Deconstructs,* 224–52. New York: Routledge.

Jao, Y.C. and Leung, C.K. 1986. *China's Special Economic Zones.* Hong Kong: Oxford University Press.

Jefferson, G.H. and Singh, I. (eds) 1999. *Enterprise Reform in China.* New York: Oxford University Press.

Jian, T., Sachs, J.D. and Warner, A.M. 1996. Trends in regional inequality in China. *China Economic Review* 7(1): 1–21.

Jiang, Q. 1991. Report on 1990 Budget Outcomes and 1991 Budget Estimates of Jiangsu Province. *Xinhua Daily,* March 22.

Jin, B. and Chen, L. (eds) 1996. *Liang'an tupo: zhongguo gongye quyu fenxi (A Regional Analysis of China's Industry).* Beijing: Economic Management Press.

Jin, S. and Zhu, R. (eds) 1993. *Zhongguo de yige xiaokang shi (A Comfortable City of China).* Nanjing: Jiangsu People's Press.

Jing, Z.H. and Tang, X.M. 1990. On the multiple investment agents in investment system reforms. *Touzi yu jianshe (Investment and Construction)* 2: 4–7, 12.

Johnson, L.C. (ed.) 1993. *Cities of Jiangnan in Late Imperial China.* Albany: State University of New York Press.

Johnston, R.J., Gregory, D. and Smith, D.M. 1994. *The Dictionary of Human Geography,* 3rd edn. Oxford: Blackwell.

JPERC (Jiangsu Provincial Economic Reforms Committee) 1988. *Shinian gaige zai Jiangsu (Ten-Years of Economic Reforms in Jiangsu).* Beijing: Reform Press.

JSB (Jiangsu Statistical Bureau) 1989. *Jiangsu sishi nian* (*Forty Years of Jiangsu*). Beijing: China Statistical Press.

JSB (Jiangsu Statistical Bureau) 1990–97. *Jiangsu tongji nianjian* (*Jiangsu Statistics Yearbook*). Beijing: China Statistical Press.

JSB (Jiangsu Statistical Bureau) 1994b. *Benxiang xiandaihua licheng* (*Towards Modernization*). Nanjing: Jiangsu Statistical Bureau.

JSB 1997b. *Jiangsu Duiwai Jingji 1990–96* (*The External Economy of Jiangsu, 1990–96*). Nanjing: Jiangsu Statistical Bureau.

Kasarda, J.D. and Crenshaw, E.M. 1991. Third World urbanization. *Annual Review of Sociology* 17: 467–501.

Kato, H. 1992. Regional development in the reform period. In Garnaut, R. and Liu, G. (eds), *Economic Reform and Internationalization*, 116–36. London: Allen and Unwin.

Khan, Z.S. 1991. *Patterns of Direct Foreign Investment in China.* Washington, DC: World Bank.

Knight, J. and Song, L. 1993. The spatial contribution to income inequality in rural China. *Cambridge Journal of Economics* 17: 195–213.

Knox, P.L. and Taylor, P.J. (eds) 1995. *World Cities in a World-System.* Cambridge: Cambridge University Press.

Kornai, J. 1980. *Economics of Shortage.* Amsterdam: North-Holland.

Kornai, J. 1992. *The Socialist System.* Princeton: Princeton University Press.

Koropeckyj, I.S. 1972. Equalization of regional development in Socialist countries. *Economic Development and Cultural Change* 21(1): 68–86.

Krebs, G. 1982. Regional inequalities during the process of national economic development. *Geoforum* 13(2): 71–82.

Krugman, P. 1991. *Geography and Trade.* Cambridge, MA: MIT Press.

Krugman, P. 1995. *Development, Geography, and Economic Theory.* Cambridge, MA: MIT Press.

KSB (Kunshan Statistical Bureau) 1993. *Kunshan shi shehui jinji tongji ziliao* (*1949–1990*) (*Social and Economic Statistics of Kunshan City* (*1949–1990*)). Kunshan: KSB.

KSB (Kunshan Statistical Bureau) 1996–98. *Kunshan tongji nianjian* (*Kunshan Statistical Yearbook*). Kunshan: KSB.

KSB (Kunshan Statistical Bureau) 1998b. A new step of Kunshan's open economy in 1997. *Jiangsu tongji* (*Jiangsu Statistics*) 5: 29–30.

Kueh, Y.Y. 1989. The Maoist legacy and China's new industrialization strategy. *China Quarterly* 119: 420–47.

Lakshmanan, T.R. and Hua, C. 1987. Regional disparities in China. *International Regional Science Review* 11(1): 97–104.

Lardy, N.R. 1975. Centralization and decentralization in China's fiscal management. *China Quarterly* 61, 23–60.

Lardy, N.R. 1978. *Economic Growth and Distribution in China.* Cambridge: Cambridge University Press.

Lardy, N.R. 1980. Regional growth and income distribution in China. In Dernberger, R.F. (ed.), *China's Development Experience in Comparative Perspective*, 153–90. Cambridge: Harvard University Press.

Lardy, N.R. 1992. *Foreign Trade and Economic Reform in China, 1978–1990.* New York: Cambridge University Press.

Lardy, N.R. 1998. *China's Unfinished Economic Revolution.* Washington, DC: Brookings Institution Press.

Larsen, K.A. 1992. *Regional Policy of China, 1949-85.* Manila: Journal of Contemporary Asia Publishers.

Lee, R. and Wills, J. (eds) 1997. *Geographies of Economies.* New York: Arnold.

Leung, C.K. 1996. Foreign manufacturing investment and regional industrial growth in Guangdong province, China. *Environment and Planning A*: 513–36.

Li, Y. and Li, S. 1986. Zhongguo sanxian jianshe de lishi zhuanzhe (Historical changes in China's Third Front construction). *Liaowang zhoukan* (*Outlook Weekly*), June 30: 10–11.

Li, W. *et al.* (eds) 1990. *Zhongguo gongye dili* (*China's Industrial Geography*). Beijing: Science Press.

Liebowitz, R.D. 1987. Soviet investment strategy. *Annals of the Association of American Geographers* 77(3): 396–407.

Lin, G.C.S. 1997. *Red Capitalism in South China.* Vancouver: UBC Press.

Linge, G.J.R. and Forbes, D.K. 1990. The space economy of China. In Linge, G.J.R. and Forbes, D.K. (eds), *China's Spatial Economy.* New York: Oxford University Press.

Liu, G. (ed.) 1984. *Zhongguo jingji fazhan zhanlue wenti yanjiu* (*A Study on China's Economic Development Strategies*). Shanghai: People's Press.

Liu, Y.L. 1992. Reform from below: the private economy and local politics in the rural industrialization of Wenzhou. *China Quarterly* 130: 293–316.

Lo, C.P. 1990. The geography of rural regional inequality in mainland China. *Transactions, Institute of British Geographers, N.S.* 15: 466–86.

Lu, D. *et al.* 1990. *Zhongguo gongye buju de lilun yu shijian* (*Theory and Practice of Industrial Location in China*). Beijing: Science Press.

Lu, D. and Sit, V. 1997. *Zhongguo quru fazhang baogao* (*Report on China's Regional Development*). Beijing: Science Press.

Lucas, R.E. 1988. On the mechanics of economic development. *Journal of Monetary Economics* 22: 3–42.

Luo, Y. 1998. *International Investment Strategies in the People's Republic of China.* England: Ashgate Publishing.

Lyons, T.P. 1985. China's cellular economy. *Journal of Comparative Economics* 9(2): 125–44.

Lyons, T.P. 1987. *Economic Integration and Planning in Maoist China.* New York: Columbia University Press.

Lyons, T.P. 1991. Interprovincial disparities in China. *Economic Development and Cultural Change* 39(3): 471–506.

Lyons, T.P. 1998. Intraprovincial disparities in China: Fujian province, 1978–1995. *Economic Geography* 74(4): 405–32.

Ma, J. 1996. China's banking sector. *Journal of Contemporary China* 5(12): 155–69.

Ma, L.J.C. and Fan, M. 1994. Urbanisation from below. *Urban Studies* 31(10): 1625–45.

Ma, L.J.C. and Noble, A.G. 1986. Chinese cities. *Urban Geography* 7(4): 279–90.

Ma, L.J.C. and Wei, Y. 1997. Determinants of state investment in China. *Tijdschrift voor Economische en Sociale Geografie* (*Journal of Economic and Social Geography*) 88(3): 211–25.

McCombie, J.S.L. 1988. A synoptic view of regional growth and unemployment. *Urban Studies* 25: 267–81.

McDaniels, I.K. 1998. Beijing rethinks investment policies. *China Business Review*, March–April: 4.

Malecki, E.J. 1991. *Technology and Economic Development.* New York: John Wiley & Sons.

Mallick, R. 1993. Convergence of state per capita incomes. *Growth and Change* 24: 321–40.

Mankiw, N.G., Romer, D. and Weil, D.N. 1992. A contribution to the empirics of economic growth. *Quarterly Journal of Economics* 107: 407–37.

Mao, Z. 1956. On the ten great relationships. In *Selected Works of Mao Tsetung*, Vol. 5, 61–83. Beijing: Foreign Languages Press, 1977.

Markusen, A.R. 1987. *Regions.* Totowa, NJ: Rowman & Littlefield.

Marme, M. 1993. Heaven on earth. In Johnson, L.C. (ed.), *Cities of Jiangnan in Late Imperial China*, 17–45. Albany: State University of New York Press.

Martin, R. 1988. The political economy of Britain's north–south divide. *Transactions, Institute of British Geographers, N.S.* 13: 389–418.

Martin, R. 1989. The growth and geographical anatomy of venture capitalism in the United Kingdom. *Regional Studies* 23(5): 289–403.

Martin, R. and Sunley, P. 1998. Slow convergence? The new endogenous growth theory and regional development. *Economic Geography* 74(3): 201–27.

Marton, A.M. 1995. Mega-urbanization in southern Jiangsu. *Chinese Environment and Development* 6(1 & 2): 9–42.

Marx, K. and Engels, F. 1959. *Basic Writings on Politics and Philosophy.* New York: Archor Books.

Marx, K. and Engels, F. 1976. *The German Ideology.* Moscow: Progress Publishers.

Massey, D. 1985. New directions in space. In Gregory, D. and Urry, J. (eds), *Social Relations and Spatial Structures*, 9–19. London: Macmillan.

Maxwell, P. and Hite, J.C. 1992. The recent divergence of regional per capita incomes. *Growth and Change* 23(1): 37–53.

Mera, K. 1977. Population concentration and regional income disparities. In Hansen, N. (ed.), *Human Settlement Systems*, 155–76. Cambridge, MA: Ballinger Publishing Company.

MFTEC (Ministry of Foreign Trade and Economic Cooperation) 1994. *Zhongguo duiwai jingji maoyi nianjian* (*China Foreign Economy and Trade Yearbook*). Beijing: Social Press.

MOF (Ministry of Finance) 1992. *Zhongguo caizheng tongji* (*1950–1991*) (*China Finance Statistics* (*1950–1991*)). Beijing: Social Press.

MOF (Ministry of Finance) 1997. *Zhongguo caizheng nianjian* (*Finance Yearbook of China*). Beijing: China Finance Magazine Press.

Moran, T.H. 1999. *Foreign Direct Investment and Development.* Washington, DC: Institute for International Economics.

Morrish, M. 1997. The living geography of China. *Geography* 82(1): 3–16.

Murphey, R. 1974. The treaty ports and China's modernization. In Elvin, M. and Skinner, G.W. (eds), *The Chinese City between Two Worlds*, 17–71. Stanford: Stanford University Press.

Murphey, R. 1976. Chinese urbanization under Mao. In Berry, B.J.L. (ed.), *Urbanization and Counterurbanization*, 311–29. Beverly Hills: Sage.

Myrdal, G. 1957. *Economic Theory and Under-Developed Regions.* London: Gerald Duckworth.

Naughton, B. 1987. The decline of central control over investment in post-Mao China. In Lampton, D.M. (ed.), *Policy Implementation in Post-Mao China*, 51–80. Berkeley and Los Angeles: University of California Press.

Naughton, B. 1988. The Third Front. *China Quarterly* 155: 351–86.

Naughton, B. 1995a. *Growing Out of the Plan.* Cambridge: Cambridge University Press.

Naughton, B. 1995b. Cities in the Chinese economic system. In Davis, D.S., Kraus, R., Naughton, B. and Perry, E.J. (eds), *Urban Space in Contemporary China*, 61–89. Cambridge: Cambridge University Press.

Nee, V. 1989. A theory of market transition. *American Sociological Review* 54: 267–82.

Nee, V. and Stark, D. with Selden, M. (eds) 1989. *Remaking the Economic Institutions of Socialism.* Stanford: Stanford University Press.

Nemes Nagy, J. 1994. Regional disparities in Hungary during the period of transition to a market economy. *GeoJournal* 32(4): 363–8.

Ni, J. and Wu, G. 1988. A structural analysis of investment sites in Jiangsu province. *Jiangsu Jingji Tantao* (*Jiangsu Economic Inquiry*) 3: 7–10.

Nolan, P. 1995. *China's Rise, Russia's Fall.* New York: St. Martin's Press.

Nolan, P. and Dong, F. 1990. *Market Forces in China.* New Jersey: Zed Books.

North, D.C. 1955. Location theory and regional economic growth. *Journal of Political Economy* 69: 319–40.

Odgaard, O. 1992. *Private Enterprises in Rural China.* Brookfield: Ashgate Publishing Company.

O hUallachain, B.O. 1985. The role of foreign direct investment in the development of regional industrial systems. *Regional Studies* 20(2): 151–62.

Oi, J.C. 1989. *State and Peasant in Contemporary China.* Los Angeles: University of California Press.

Oi, J.C. 1992. Fiscal reform and the economic foundations of local state corporatism in China. *World Politics* 45(1): 99–126.

Oksenberg, M. and Tong, J. 1991. The evolution of central–provincial fiscal relations in China, 1971–1984. *China Quarterly* 128: 1–32.

Ozornoy, G.L. 1991. Some issues of regional inequality in the USSR under Gorbachev. *Regional Studies* 25(5): 381–93.

Paine, S. 1981. Spatial aspects of Chinese development. *Journal of Development Studies* 17(2): 132–95.

Pannell, C.W. 1988. Regional shifts in China's industrial output. *Professional Geographer* 40(1): 19–32.

Pannell, C.W. and Ma, L.J.C. 1983. *China.* New York: John Wiley & Sons.

Park, J. 1997. *The Special Economic Zones of China and Their Impact on Its Economic Development.* Westport, Conn: Praeger.

Parsons, D.W. 1988. *The Political Economy of British Regional Policy.* New York: Routledge.

Peng, G.L. and Huang, S.T. (eds) 1992. *Zhongguo gaige quanshu: touzi tizhi gaige juang* (*Encyclopedia on China's Reforms: Investment System Reforms*). Dalian: Dalian Press.

Peng, M., Lu, B. and Wang, D. (eds) 1989. *Dangdai zhongguo de jiben jianshe* (*Basic Capital Construction Investment in Contemporary China*). Beijing: China Social Science Press.

People's Government of Jiangsu Province 1997. *Jiangsu nianjian* (*Jiangsu Almanac*). Nanjing: Jiangsu Almanac House.

Perroux, F. 1950. Economic space. *Quarterly Journal of Economics* 64: 89–104.

Perroux, F. 1955. Note sur la notion de 'pole de croissance'. In Mckee, D.L., Dean, R.D. and Leahy, W.H. (eds), *Regional Economics*, 93–104, 1970. New York: Free Press.

Pickles, J. and Smith, A. (eds) 1998. *Theorizing Transition.* New York: Routledge.

Piore, M.J. and Sabel, C.F. 1984. *The Second Industrial Divide*. New York: Basic Books.

Plestina, D. 1992. *Regional Development in Communist Yugoslavia*. Boulder: Westview Press.

Pomfret, R. 1989. *Equity Joint Ventures in Jiangsu Province*. New York: Longman.

Pomfret, R. 1995. Taiwan's involvement in Jiangsu province. In La Croix, S.J., Plumner, M. and Lee, K. (eds), *Emerging Patterns of East Asian Investment in China*, 167–78. New York: M.E. Sharpe.

Porter, M. 1990. *The Competitive Advantage of Nations*. New York: Free Press.

Potter, P.B. 1995. Foreign investment law in the People's Republic of China. *China Quarterly* 141: 155–85.

Prime, P.B. 1992. Industry's response to market liberalization in China. *Economic Development and Cultural Change* 41(1): 27–50.

Project Group, State Council Research Department. 1994. *Woguo suoyouzhi jiegou biange de qushi he duice* (*Trends and Strategies of Reforming China's Ownership Structure*). Beijing: China Economic Press.

Qu, T. and Green, M.B. 1997. *Chinese Foreign Direct Investment*. Aldershot: Ashgate.

Raiser, M. 1998. Subsidising inequality: economic reforms, fiscal transfers and convergence across Chinese provinces. *Journal of Development Studies* 34(3): 1–26.

Rich, DC and Linge, G.J.R. 1991. *The State and the Spatial Management of Industrial Change*. New York: Routledge.

Richardson, H. 1978. *Regional Economics*. Urbana, IL: University of Illinois Press.

Riskin, C. 1987. *China's Political Economy*. Oxford: Oxford University Press.

Romer, P. 1986. Increasing returns and long-run growth. *Journal of Political Economy* 92: 1002–37.

Rozelle, S. 1994. Rural industrialization and increasing inequality. *Journal of Comparative Economics* 19(4): 362–91.

Rozelle, S. 1996. Stagnation without equity. *China Journal* 35: 63–92.

Sabin, L. 1994. New bosses in the workers' state: the growth of non-state sector employment in China. *China Quarterly* 140: 944–70.

Schiffer, J.R. 1985. Interpretations of the issue of 'inequality' in Soviet regional policy debates. *International Journal of Urban and Regional Research* 9(4): 508–32.

Schiffer, J.R. 1989. *Soviet Regional Economic Policy*. London: Macmillan.

Schueller, M. 1997. Liaoning: struggling with the burdens of the past. In Goodman, D.S.G. (ed.), *China's Provinces in Reform*. New York: Routledge.

Scott, A.J. 1988. *New Industrial Space*. London: Pion.

Shen, J. 1993. The influence of fiscal systems on industrial structural adjustment in Jiangsu. *Jiangsu Jingji Tantao* (*Jiangsu Economic Inquiry*). 4: 32–40.

Shih, J.C. 1992. *Chinese Rural Society in Transition*. Berkeley: Institute of East Asian Studies.

Shirk, S.L. 1993. *The Political Logic of Economic Reform in China*. Berkeley and Los Angeles: University of California Press.

Shue, V. 1988. *The Reach of the State*. Stanford: Stanford University Press.

Sidaway, J.D. and Simon, D. 1990. Spatial policies and uneven development in the 'Marxist–Leninist' states of the Third World. In Simon, D. (ed.), *Third World Regional Development*. London: Paul Chapman.

Simmons, A.B. 1979. Slowing metropolitan city growth in Asia. *Population and Development Review* 5(1): 87–104.

Singh, I. 1992. *China: Industrial Policies for an Economy in Transition*. Washington, DC: World Bank. Discussion Papers, No. 143.

Siriopoulos, C. and Asteriou, D. 1998. Testing for convergence across the Greek regions. *Regional Studies* 32(6): 537–46.

Skinner, G.W. 1977. Regional urbanization in nineteenth-century China. In Skinner, G.W. (ed.), *The City in Late Imperial China*, 211–49. Stanford: Stanford University Press.

Slater, D. 1975. Underdevelopment and spatial inequality. *Progress in Planning* 4: 97–167.

Smith, A. 1998. *Reconstructing the Regional Economy: Industrial Transformation and Regional Development in Slovakia*. Massachusetts: Edward Elgar Publishing.

Smith, D. 1975. Neoclassical growth models and regional growth in the US. *Journal of Regional Science* 15: 165–81.

Smith, D.M. 1995. *Geography and Social Justice*. Oxford: Blackwell.

Smith, N. 1984. *Uneven Development*. London: Basil Blackwell.

Soja, E. 1980. The socio-spatial dialectic. *Annals of the Association of American Geographers* 70(2): 207–25.

Solinger, D.J. 1996. Despite decentralization. *China Quarterly* 145: 1–34.

Song, L. *et al.* (eds) 1998. *Jiangsu jingji shehui xingshi fenxi yu yuce* (*Analysis and Prediction of Jiangsu's Economic and Social Conditions*). Nanjing: Jiangsu People's Press.

SPC (State Planning Commission) 1996–97. *Zhongguo touzi baipishu* (*China Investment White Report*). Beijing: International Culture Press.

SSB (State Statistical Bureau) 1986. Economic growth in different areas. *Beijing Review* 29(49): 21–4.

SSB (State Statistical Bureau) 1990. *Quanguo gesheng zizhiqu zhiziashi lishi tongji ziliao huibian* (*1949–89*) (*A Compilation of Historical Statistical Materials of China's Provinces, Autonomous Regions, and Centrally Administered Municipalities* (*1949–1989*)). Beijing: China Statistical Press.

SSB (State Statistical Bureau) 1991–97. *Zhongguo tongji nianjian* (*Statistical Yearbook of China*). Beijing: China Statistical Press.

SSB (State Statistical Bureau) 1991b. *Zhongguo zhongdian jianshe* (*China's Key-Point Construction*). Beijing: Law Press.

SSB (State Statistical Bureau) 1995b–97b. *Zhongguo chengshi tongji nianjian* (*Urban Statistical Yearbook of China*). Beijing: China Statistical Press.

SSB (State Statistical Bureau) 1996c. *Gaige kaifang shiqi nian de zhongguo diqu jingji* (*China's Regional Economy: A Profile of 17 Years of Reforms and Opening-up*). Beijing: China Statistical Press.

SSB (State Statistical Bureau) 1997c. *Zhongguo guding zichan touzi tongji nianjian 1950–1995* (*Statistical Yearbook of Fixed Asset Investment in China*). Beijing: China Statistical Press.

State Council 1956. *First Five-Year Plan for the Development of the National Economy of the People's Republic of China in 1953–1957*. Beijing: Foreign Language Press.

State Council 1986. *Zhonghua renmin gongheguo guomin jingji he shehui fazhan diqige wunian jihua 1986–1990* (*The Seventh Five-Year Plan for National Economic and Social Developments in the People's Republic of China 1986–1990*). Beijing: People's Press.

Steinfeld, E.S. 1998. *Forging Reform in China*. New York: Cambridge University Press.

Stohr, W.B. and Taylor, D.R.F. (eds) 1981. *Development from Above or Below?* New York: John Wiley & Sons.

Stohr, W.B. and Todtling, F. 1977. Spatial equality. *Papers of the Regional Science Association* 38: 33–53.

Storper, M. 1997. *The Regional World.* New York: The Guilford Press.

Storper, M. and Walker, R. 1989. *The Capitalist Imperative.* New York: Basil Blackwell.

Su, S. and Veeck, G. 1995. Foreign capital acquisition by rural industries in Jiangsu. *Chinese Environment and Development* 6(1 & 2): 169–92.

Sun, H. 1998. *Foreign Investment and Economic Development in China.* Brookfield: Ashgate.

Sun, J. 1988. Some thoughts on deepening the reform of investment systems. *Guangli shijie (Management World)* 5: 44–48.

Szelenyi, I. 1981. Structural changes of and alternatives to capitalist development in the contemporary urban and regional system. *International Journal of Urban and Regional Research* 5(1): 1–14.

Tabuchi, T. 1988. Interregional income differentials and migration. *Regional Studies* 22(1): 1–10.

Tang, J.Z. and Ma, L.J.C. 1985. Evolution of urban collective enterprises in China. *China Quarterly* 104: 614–40.

Taylor, M. and Thrift, N. 1982. *The Geography of Multinationals.* New York: St. Martin's Press.

Taylor, P.J. 1993. *Political Geography.* Essex: Longman.

Thrift, N.J. and Forbes, D.K. 1986. *The Price of War.* London: Allen & Unwin.

Tian, G. 1996. *Shanghai's Role in the Economic Development of China.* London: Praeger.

Tong, J. 1991. Central–provincial fiscal relations in post-Mao China. Department of Political Science, UCLA. Manuscript.

Tsui, K.Y. 1991. China's regional inequality, 1952–1985. *Journal of Comparative Economics* 15: 1–21.

Turnock, D. 1984. Postwar studies on the human geography of Eastern Europe. *Progress in Human Geography* 8: 315–46.

Veeck, G. (ed.) 1991a. *The Uneven Landscape.* Baton Rouge, La.: Geoscience Publications, Dept. of Geography and Anthropology, Louisiana State University.

Veeck, G. 1991b. Regional variations in employment and income in Jiangsu province. In Ginsburg, N., Koppol, B. and McGee, T.G. (eds), *The Extended Metropolis,* 157–76. Honolulu: University of Hawaii Press.

Veeck, G. and Pannell, C.P. 1989. Rural economic restructuring and farm household income in Jiangsu, People's Republic of China. *Annals of the Association of American Geographers* 79(2): 275–92.

Vogel, E.F. 1989. *One Step Ahead in China.* Cambridge: Harvard University Press.

Walder, A.G. (ed.) 1995a. *The Waning of the Communist State.* Berkeley and Los Angeles: University of California Press.

Walder, A.G. 1995b. Local governments as industrial firms. *American Journal of Sociology* 101(2): 263–301.

Wang, H. 1986. *Xin zhongguo gongye jingjishi (A History of New China's Industrial Economy).* Beijing: Economic Management Press.

Wang, L. (ed.) 1997. *Zhongguo waishang touzi baogao (Report on Foreign Investment in China).* Beijing: Economic Management Press.

Wang, S. and Hu, A. 1993. *Zhongguo guojia nengli baogao* (*Report on China's State Capacity*). Shenyang: Liaoning People's Press.

Wang, X. 1990. Shilun woguo sanxian gongye de tiaozheng gaizao (On the Adjustment and Transformation of China's Third Front Industries). *Shehui kexue yanjiu* (*Social Science Research*) 3: 33–37.

Wang, Y., Wu, Y., Yang, J. and Wang, Q. (eds) 1998. *Zhongguo Quyu Jingji Zhengce Yanjiu* (*China's Regional Economic Policy Studies*). Beijing: China Planning Press.

Wang, Z. 1991. The 'Shanghai phenomenon' in China's economic growth. *Shanghai gaige* (*Shanghai Reforms*) 5: 15–17; 6: 14–15; 7: 10–11.

Wedeman, A. 1993. *Bamboo Walls and Brick Ramparts*. Ph.D. Dissertation, UCLA.

Wedeman, A. 1999. Agency and fiscal dependence in central–provincial relations in China. *Journal of Contemporary China* 8(20): 103–22.

Wei, H. 1995. *Quyu jingji fazhang de xin geju* (*New Patterns of Regional Economic Development in China*). Kunming: Yunnan People's Press.

Wei, H., Liu, K., Zhou, M., Yang, D. and Hu, W. 1997. *Zhongguo diqu fazhan* (*China's Regional Development*). Beijing: Economic Management Press.

Wei, Y. 1989. Urbanization and its regional economic mechanism in Zhejiang province. *Zhejiang xuekan* (*Zhejiang Journal*) 4: 85–7.

Wei, Y. 1993. China's regional development studies: major topics and recent progress. *Jinji dili* (*Economic Geography*) 13(4): 1–7.

Wei, Y. 1994. Economic policy, urban policy, and the growth of large cities in China. *Habitat International* 18(4): 53–65.

Wei, Y. 1995. Spatial and temporal variations of the relationship between state investment and industrial output in China. *Tijdschrift voor Economische en Sociale Geografie* (*Journal of Economic and Social Geography*) 86(2): 129–36.

Wei, Y. 1996. Fiscal systems and uneven regional development in China, 1978–1991. *Geoforum* 27(3): 329–44.

Wei, Y. 1997a. *Patterns of Regional Inequality in China: Globalization, the State, and Localities*. Ph.D. Dissertation, UCLA.

Wei, Y. 1997b. Interregional migration in transitional socialist countries: the case of China. *GeoJournal* 41(3): 205–14.

Wei, Y. 1998. Regional inequality of industrial output in China, 1952–90. *Geografiska Annaler B* 80(1): 1–15.

Wei, Y. 1999a. Regional inequality in China. *Progress in Human Geography* 23(1): 48–58.

Wei, Y. 1999b. Changes in industrial ownership structure in China. *Geography* 84(3): 193–203.

Wei, Y. and Dutt, A.K. 1992. Distribution of medical facilities in the People's Republic of China with particular reference to rural areas. *Indian Geographical Journal* 67(1): 9–19.

Wei, Y. and Fan, C.C. 1997. Intraprovincial inequality in China: a case study of Jiangsu. Paper presented at the annual meeting of the Association of American Geographers, Fort Worth, Texas, April 1–5 (forthcoming in *Professional Geographer*).

Wei, Y. and Ma, L.J.C. 1996. Changing patterns of spatial inequality in China. *Third World Planning Review* 18(2): 177–91.

Weitzman, M.L. and Xu, C. 1994. Chinese township–village enterprises as vaguely defined cooperatives. *Journal of Comparative Economics* 18(2): 121–45.

Weng, Q. 1998. Local impacts of the post-Mao development strategy: the case of the Zhujiang Delta, southern China. *International Journal of Urban and Regional Research* 22 (3): 425–42.

White, L.T., III. 1989. *Shanghai Shanghaied*? Hong Kong: Centre of Asian Studies, University of Hong Kong.

Williamson, J.G. 1965. Regional inequality and the process of national development. *Economic Development and Cultural Change* 13(2): 3–83.

Wong, C.P.W. 1991. Central–local relations in an era of fiscal decline. *China Quarterly* 128: 691–715.

Wong, C.P.W. 1997. Overview of issues in local public finance in the PRC. In Wong, C. (ed.), *Financing Local Government in the People's Republic of China*. Hong Kong: Oxford University Press.

Wong, C.P.W., Heady, C. and Woo, W.T. 1995. *Fiscal Management and Economic Reform in the People's Republic of China*. Hong Kong: Oxford University Press.

Woo, W.T., Parker, S. and Sachs, J.D. (eds) 1997. *Economies in Transition*. Cambridge: MIT Press.

World Bank 1985. *China: Long-term Development Issues and Options*. Washington, DC: World Bank.

World Bank 1990. *China: Macroeconomic Strategy and Industrial Growth under Decentralized Socialism*. Washington, DC: World Bank.

World Bank 1992a. *China: Reform and the Role of the Plan in the 1990s*. Washington, DC: World Bank.

World Bank 1992b. *China: Strategy for Reducing Poverty in the 1990s*. Washington, DC: World Bank.

World Bank 1997. *Sharing Rising Incomes*. Washington, DC: World Bank.

Wu, C. 1987. Chinese socialism and uneven development. In Forbes, D. and Thrift, N. (eds), *The Socialist Third World*, 53–97. New York: Basil Blackwell.

Wu, Y. 1967. *The Spatial Economy of Communist China*. New York: Praeger.

Wuxi, the People's Government 1996. *Maixiang 21 shiji de wuxi (Wuxi: Towards the 21st Century)*. Beijing: China Yanshi Press.

Xia, T. and Li, Z. 1993. *Zhongguo touzi jianshi* (*A Brief History of China's Investment*). Beijing: China Financial Economy Press.

Xiang, H., Jiang, W. and Li, J. (eds) 1992. *Zhongguo gaige quanshu: Caizheng gaige juang* (*Encyclopedia on China's Reforms: Financial Reforms*). Dalian: Dalian Press.

Xie, Y. and Costa, F.J. 1991. The impact of economic reforms on the urban economy of the People's Republic of China. *Professional Geographer* 43(3): 318–35.

Xue, M. 1982. *Current Economic Problems in China* (edited, translated, and with an introduction by K.K. Fung). Boulder: Westview Press.

Yang, D. 1990. Patterns of China's regional development strategy. *China Quarterly* 122: 230–57.

Yang, D. 1997. *Beyond Beijing*. New York: Routledge.

Yang, K. 1993. *Maixiang kongjian yitihua* (*Towards Spatial Integration*). Chengdu: Sichuan People's Press.

Yang, W. 1992. An empirical analysis of the change of regional income inequality. *Jinji yanjiu* (*Economic Research*) 1: 70–4, 34.

Yenal, O. 1990. Chinese reforms, inflation and the allocation of investment in a socialist economy. *World Development* 18(5): 707–21.

Yeung, H.W. 1994. Hong Kong firms in the ASEAN region. *Environment and Planning A* 26: 1931–56.

Yusuf, S. and Wu, W. 1997. *The Dynamics of Urban Growth in Three Chinese Cities.* New York: Oxford University Press, published for the World Bank.

Zhang, F. and Wei, Y. 1986. *Yangzhou shiyu shehui jingji fazhuan yu chengzheng tixi buju yangjiu baogao (A Study of Socioeconomic Development and Urban Systems Allocation in the City-Region of Yangzhou).* Monograph. Nanjing: Nanjing Institute of Geography, Chinese Academy of Sciences.

Zhang, L. 1994. Location-specific advantages and manufacturing direct foreign investment in south China. *World Development* 22(1): 45–53.

Zhang, S. (ed.) 1994. *Kunshan kaifa zidao (The Development Path of Kunshan).* Nanjing: Jiangsu People's Press.

Zhang, S. 1996. *Kunshan fazhan guiji jishi (Portrait of the Development Path of Kunshan).* Nanjing: Jiangsu People's Press.

Zhang, T. and Su, S. (eds) 1993. *Jiangsu sheng jingji dili (Economic Geography of Jiangsu).* Beijing: Xinhua Press.

Zhao, S. 1994. *Geography of China.* New York: John Wiley & Sons.

Zhao, X. 1996. Spatial disparities and economic development in China, 1953–92. *Development and Change* 27: 131–63.

Zhao, X. and Zhang, L. 1995. Urban performance and the control of urban size in China. *Urban Studies* 32(4–5): 813–45.

Zheng, Y. 1995. *Institutional Change, Local Developmentalism, and Economic Growth: The Making of Semi-Federalism in Reform China,* Ph.D. dissertation. Princeton University.

Zhou, D. and Zhou, H. (eds) 1997. *Suzhou he Wenzhou fazhen bijiao yanjiu (A Comparative Study on Development of Suzhou and Wenzhou).* Suzhou: Suzhou University Press.

Zhou, Q. 1996. Interprovincial resource transfers in China, 1952–90. *International Journal of Urban and Regional Research* 20(4): 571–86.

Zhou, R.L. and Zhang, R.L. (eds) 1994. *Zhongguo chengxiang xietiao fazuan yanjiu (A Study of Urban and Rural Integrated Development in China).* Oxford: Oxford University Press.

Index